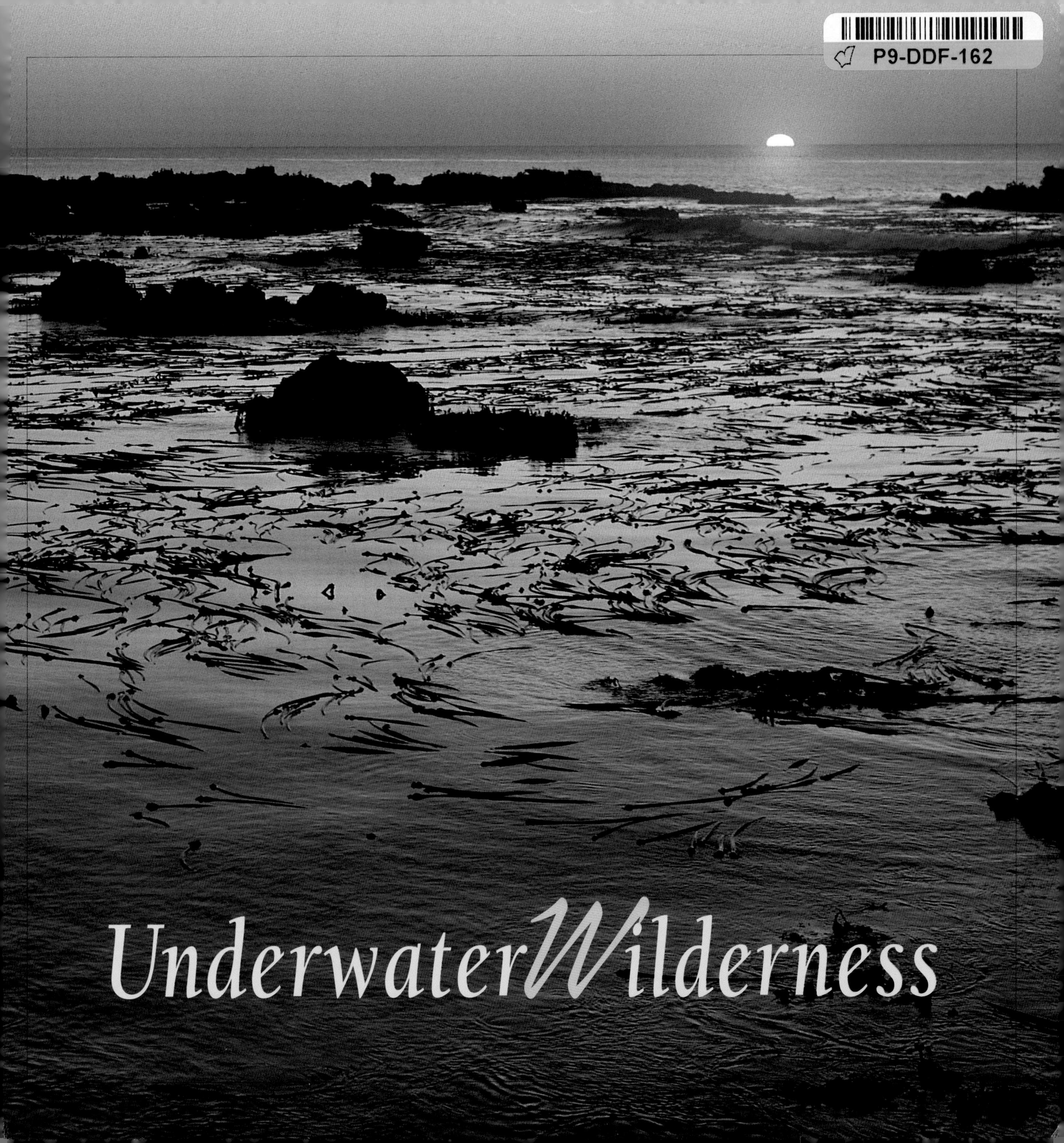

Underwater Wilderness

Editor: Nora L. Deans
Design: Ann W. Douden

Published in the United States of America by
ROBERTS RINEHART PUBLISHERS
5455 Spine Road, Mezzanine West
Boulder, Colorado 80301
Tel. (303) 530-4400

Published in the UK and Ireland by
ROBERTS RINEHART PUBLISHERS
Trinity House, Charleston Road
Dublin 6, Ireland

Distributed in the U.S. and Canada by
Publishers Group West

Printed in Hong Kong by Palace Press

Library of Congress Catalog
Card Number: 96-068565

International Standard Book Number:
1-57098-104-3

All photographs by Charles Seaborn except photographs by Jesse Cancelmo, pg. 91, 94, 99, back cover marble grouper; Norman Depres, pg. 17, 25; Pete Nawrocky, pg. 31, 40, 49, 55; Nora Seaborn, pg. 65, 135, cover flap; and Herb Segars, pg. 14, 15, 18, 20, 21, 22, 23, 24, 26, 28, 34, 35, 39, 44, 50, 57, back cover anemone shrimp and northern seahorse.

MONTEREY BAY
AQUARIUM

ROBERTS
RINEHART
PUBLISHERS

Moon jellies drift throughout the world's oceans.

Title page: Sunset on the southern Oregon coast.

DEDICATION

TO NORA
FOR FILLING IN
THE SPACES

Underwater Wilderness

Life in America's National Marine Sanctuaries and Reserves

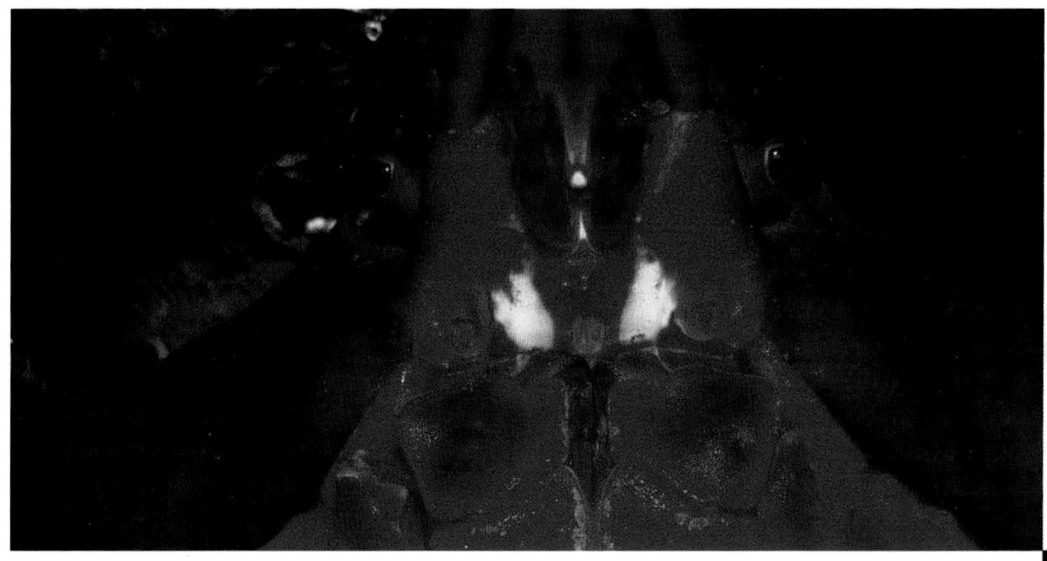

Face to face with a northern kelp crab in Puget Sound, Washington.

BY

Charles Seaborn

ROBERTS RINEHART PUBLISHERS
&
THE MONTEREY BAY AQUARIUM

A queen parrotfish rests in its nightly coccoon.

Feather duster worms are conspicuous, but shy, inhabitants of Caribbean coral reefs.

Preface

This book is a "snapshot" of the incredibly diverse marine life in the 33 designated National Marine Sanctuaries and Reserves along the coast of this country.

This tour of our coast is divided into seven distinct regions based on the kinds of marine life you find in each of these "zones." We begin in Maine, travel down the eastern seaboard, through the Florida Keys, into the Gulf. On the Pacific side the tour picks up in southern California, moving up the West Coast to Alaska, then jumping to Hawaii and American Samoa.

Many people assisted me in the development of this book, and I am most grateful to all of them. To name a few, my thanks to Dr. Eric Lindgren for his thoughtful comments during the conceptual phase of the project. Brian O'Callaghan and Susan Jurasz of Sea Reach Limited provided support in a variety of areas, including numerous late-night research and brainstorming sessions. Bob Hollis at Oceanic gave invaluable assistance through the generous use of their excellent diving equipment for much of the field work. John de Boeck was kind enough to provide sea time aboard the *Clavella* in the Queen Charlotte Islands. Mike Kennedy at Cape Henlopen State Park, Delaware, gave me both professional and personal insights into the wonders of northeast marine life.

I would also like to thank the staff at the Sanctuaries and Reserves Division of the National Oceanic and Atmospheric Administration, including Charlie Wahle, Maureen Wilmot, Joyce Atkinson, Harriet Sopher and Brady Phillips at the national headquarters, as well as Reed Bohne, Edward Bowlby, Dr. Stephen Gittings and Ben Haskell, who reviewed the chapters for scientific accuracy. Thoughtful comments were also provided by Dr. Steven Webster and Dave Powell of the Monterey Bay Aquarium. My good friend and diving companion Nora "Remora" Deans gave it the "value-added" that represents the quality of the Monterey Bay Aquarium.

This book became a reality thanks to the talented staff at Roberts Rinehart Publishers. Betsy Armstrong navigated it through the editorial phase and Ann Douden designed a stunning presentation. My sincere thanks to Rick Rinehart and Jack Van Zandt for simply making it happen.

CONTENTS

Male goldfish "yawning" on a tropical Pacific coral reef.

Underwater Wilderness

Contents

Hawksbill sea turtle swimming against a Hawaiian sky.

INTRODUCTION 8

CHAPTER 1 15
North Atlantic Coast

CHAPTER 2 35
The South Atlantic Coast

CHAPTER 3 61
The Florida Keys

CHAPTER 4 85
The Gulf of Mexico

CHAPTER 5 105
The California Coast

CHAPTER 6 135
The Pacific Northwest and Alaska

CHAPTER 7 159
Hawaii and American Samoa

GLOSSARY 189

LISTING 191
National Marine Sanctuaries and National Estuarine Research Reserves

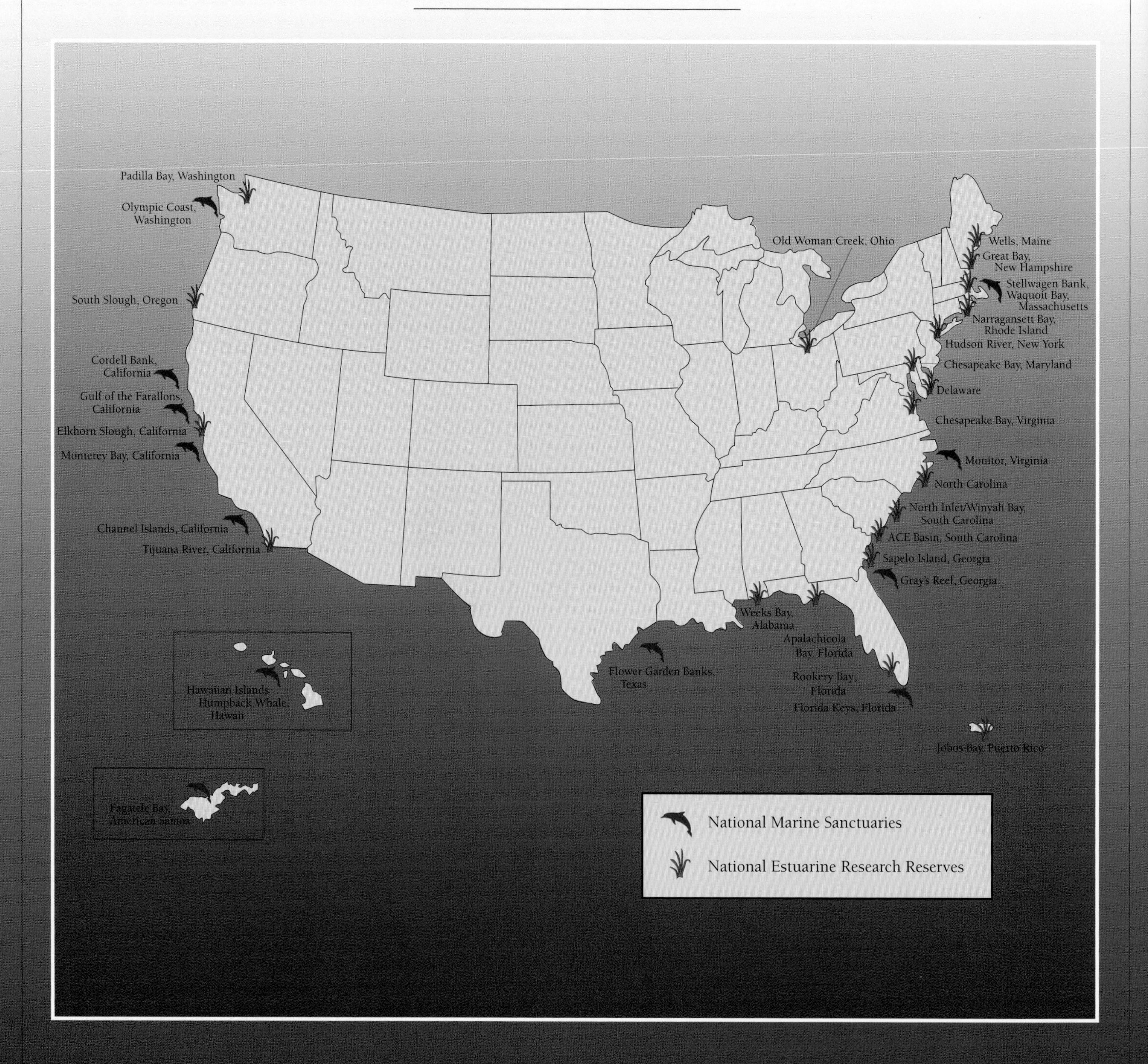

Padilla Bay, Washington
Olympic Coast, Washington
South Slough, Oregon
Cordell Bank, California
Gulf of the Farallons, California
Elkhorn Slough, California
Monterey Bay, California
Channel Islands, California
Tijuana River, California
Old Woman Creek, Ohio
Wells, Maine
Great Bay, New Hampshire
Stellwagen Bank, Waquoit Bay, Massachusetts
Narragansett Bay, Rhode Island
Hudson River, New York
Chesapeake Bay, Maryland
Delaware
Chesapeake Bay, Virginia
Monitor, Virginia
North Carolina
North Inlet/Winyah Bay, South Carolina
ACE Basin, South Carolina
Sapelo Island, Georgia
Gray's Reef, Georgia
Weeks Bay, Alabama
Apalachicola Bay, Florida
Flower Garden Banks, Texas
Rookery Bay, Florida
Florida Keys, Florida
Jobos Bay, Puerto Rico
Hawaiian Islands Humpback Whale, Hawaii
Fagatele Bay, American Samoa
National Marine Sanctuaries
National Estuarine Research Reserves

IntroduCtion

Coral reef scene, Indo-Pacific.

Fifteen years after Congress declared 1980 the "Year of the Coast," it is safe to say that the United States has truly become a nation of oceans. The increased interest in our coastal areas throughout the country reveals itself in a variety of ways. First the good news: public awareness of ocean and coastal issues has increased greatly, particularly with respect to their value as natural resources. Annual attendance at public aquariums has never been higher and it seems that every coastal city has plans to build one. The number of recreational scuba divers has been increasing steadily, as has "adventure travel" for the diving enthusiast to remote locals such as New Guinea and Fiji.

The news, however, is not all good. Interest in the oceans and the health of our coast has also been heightened by more disturbing events. Even prior to the 1988 Exxon *Valdez* oil spill disaster in Alaska, articles about coastal pollution and ocean conservation began to appear in such unlikely publications as *Business Week* and *The Atlantic*.

As the world becomes smaller and smaller as a result of a growing population

Bat stars are a ubiquitous species of the California rocky reef habitat, and they come in all colors. This orange-red variety is quite common, as are the orange cup corals surrounding it.

and diminishing natural resources, the oceans will become increasingly important to us. Americans look more and more to the coast for commerce, food, defense and recreation. Already over half of our population lives within 50 miles of the roughly 90,000 miles of American shoreline. With the extension of United States jurisdiction out to 200 miles offshore, over 2.2 million square miles of ocean lies under our direct economic control.

The magnitude of this kind of responsibility is difficult to imagine even for those of us who directly manage ocean and coastal resources. Most of us have experienced the ocean on a more personal level. All of us who have lived along the coast or who have enjoyed a summer vacation at the beach remember a favorite time, place or event that happened along the edge of the sea. Many may recall a quiet walk at sunset along a stretch of deserted beach, with waves lapping gently (or fiercely) at their feet. Others may remember a more adventurous experience, such as hiking along a rocky, mountainous shore with waves crashing on cliffs below.

For myself, one experience along the shore stands out among many, and it combines the serenity and reflection of a quiet beach walk with the adventure of coastal rock climbing. At dusk on a summer's day, two friends and I reached the narrow beach of Shell Island, a small, rocky outcrop half a mile off the shore of North Cove, on Cape Arago in southern Oregon. We were greeted by a loud, not-so-friendly-sounding chorus of the Steller sea lions. Pulling the raft up on the beach, we separated to explore what we perceived to be "our" island. Frankly, I was exhausted from the trip, so I lay down on the beach, half of my drysuit-clad body in the water and half out. On a tropical beach this would not have been so odd, but in the chilly waters of the Pacific Northwest it hit me that I was warm, the sun was out, and it seemed like I was in a very different kind of place than the Oregon coast.

I looked down, and discovered how this fledgling island got its name. There was no sand on the "beach"—just shells, in all shapes, colors and sizes. Most, however, were very small. I spent the next twenty minutes in perfect comfort, as the sun set and small waves lapped at my drysuited body, while I sorted through handful after handful of sea shells. At first I was impressed by the diversity of shells that each new handful revealed. This brought on an intense, but brief, desire to collect as many different kinds as I could. Then I asked myself: How did they get here? Why here? What is it about this particular spot on the coast that caused these shells to accumulate? What combination of waves, tides and currents allowed this to happen? How many years has this been going on? And, finally, does anyone else besides myself and the sea lions know about it?

I couldn't wait to tell my companions about my discovery. As we half-swam, half-paddled back to shore we began to compare notes on our brief, but memorable, experiences on the island. Each had an exciting story to tell of the adventure that always comes with explorations along the shore, one of the reasons we are all so taken by spending time at the beach.

The character of the United States coast varies greatly from shore to shore. The windswept dunes on the barrier islands of North Carolina differ greatly from the coral reefs off the Florida Keys. The volcanic, rocky shores of Hawaii stand in sharp contrast to the shell beach I explored off the southern coast of Oregon. And while I still remember that experience as clearly today as when it happened nearly fourteen years ago, I still want to know what made that part of the coast unique and how it differs from other coasts.

Clearly, the primary characteristics of a coast are determined by tectonic activity. The major differences between the Atlantic and Pacific coasts are due to coastal tectonic processes. Regional differences along both coasts are due to more specific, localized geologic processes such as glaciers, rivers, waves, tides and currents. The creation and distribution of bays, cliffs and beaches along both coasts is a result of these local geological events superimposed on top of the tectonic imprint.

While we are more familiar with the effects of glaciers and rivers on land, they have also played a role in the evolution of coastal and subtidal areas. The erosive action of rivers and glaciers carved deep valleys that were flooded by a rise in sea level. Bays, fjords, inlets and estuaries formed as the result of eroding forces. Other coastal features, such as deltas and beaches, were created by the accumulation of the sediments these agents brought to the coasts from the land.

On a daily basis, waves, tides and currents play a dominant role in shaping the character of a coast. Waves are constantly eroding cliffs, or during the summer months piling sand onto beaches, only to take it away again during winter storms. Tides raise and lower sea level twice or four times a day, creating a unique and environmentally demanding place for the animals and plants that live in this zone of half-water, half-land. Longshore currents also change strength and direction seasonally in many parts of our coast, building sand spits and filling in harbors.

The activities of living plants and animals can also have a great effect on the development of a coast. Beds of mussels build coastal areas, while rock-boring sea urchins aid in erosion. In Maine, Washington, Oregon and other temperate regions, animals such as barnacles and mussels live attached to the rocky shoreline. The actual features of these coasts, however, were created primarily by geologic processes. In the tropics, animals such as corals, sponges and even tube-building worms create substantial coastal features in the form of reefs. In the Pacific, many islands are constructed of reefs or atolls built entirely from the skeletal structures of such organisms, which have been going about their invertebrate business for thousands of years.

This book only begins to describe the abundance and diversity of marine life living within the political boundaries that define the United States. From the cold-temperate waters of the Pacific Northwest to the tropical waters of Hawaii and the Florida Keys, America's coastal marine life is a study in marine life distribution and adaptation. The plants and animals found here know no political boundaries. Instead, their lives are determined by natural physical barriers created by the shape of the shore, sea floor surface, water currents, tides, temperature and salinity, to name a few. Despite rapid advances in marine research over the past decade, we are just beginning to develop an understanding of how much we don't know about these complex communities of oceanic life. A recent coral reef study, for example, indicates that we may have identified only about 15 percent of the organisms that make up a "typical" section of Caribbean coral reef.

It is becoming clear that maintaining marine biodiversity is critical to the ongoing health of coastal marine environments. We simply do not know what will happen if a "critical mass" of species within a community are lost. At the same time, the recruitment of new marine scientists—particularly in the field of taxonomy—is decreasing. While the need to broaden our knowledge base increases, the national resources to do so appear to be decreasing. Programs such as the National Marine Sanctuaries and National Estuarine Research Reserves system, which set aside distinct coastal areas in perpetuity for future research, are critical in furthering our understanding of our nation's wetlands and marine resources.

It is also important that we inspire and nurture future generations of marine scientists. Institutions such as public aquariums and oceanariums can play a critical role in providing informal education opportunities for young people that provide not only solid information about our coast but excitement, encouragement and interest in marine biology, coastal geology, oceanography and other related professions. These are the people who will become the stewards of America's coasts, either as concerned citizens or by taking an active role in maintaining its natural health.

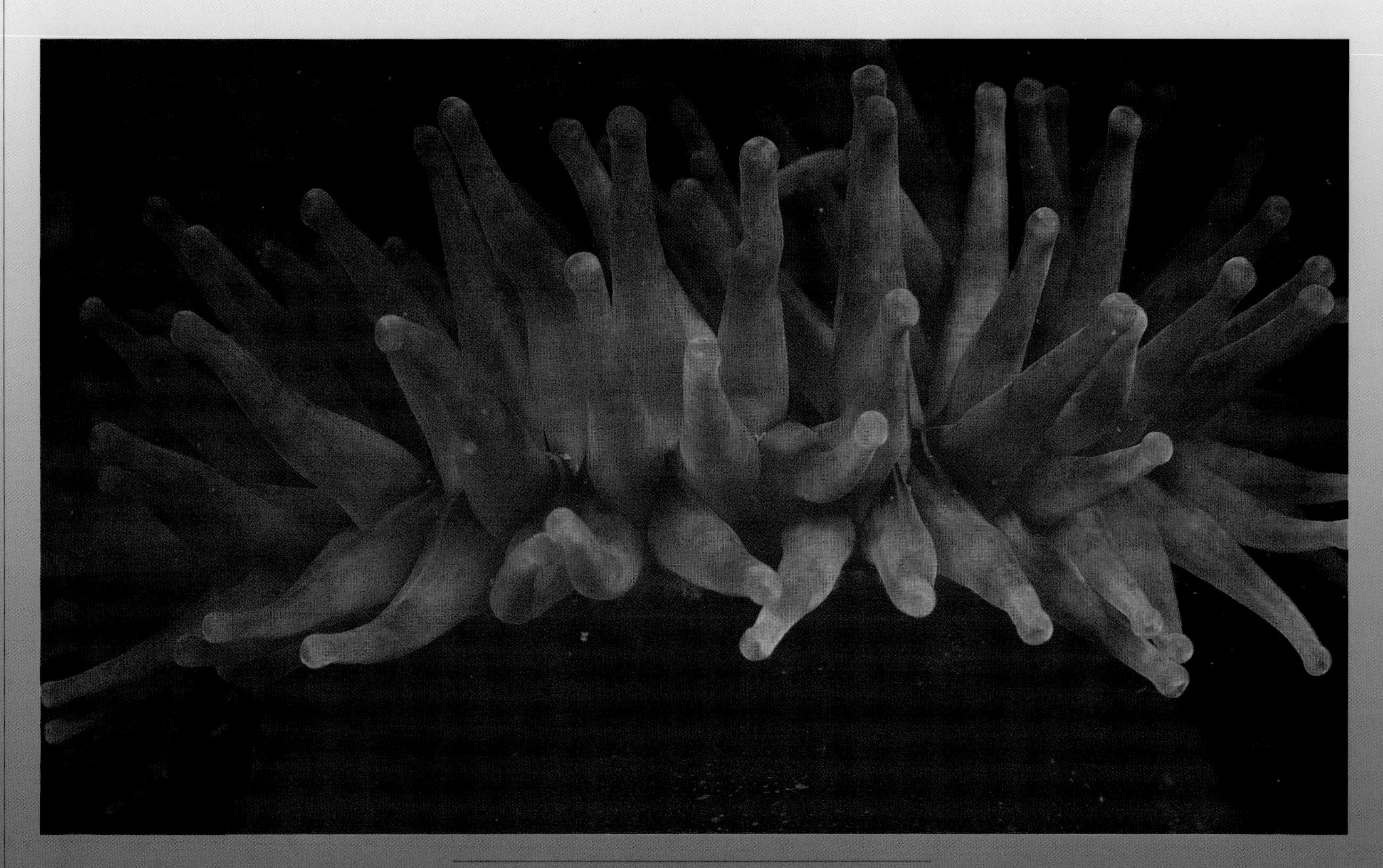

Anemones in the genus Tealia *are some of the most colorful northeastern marine animals. Related to jellyfish, they too have stinging cells in their tentacles, which are used for both feeding and defense.*

ONE

The North Atlantic Coast

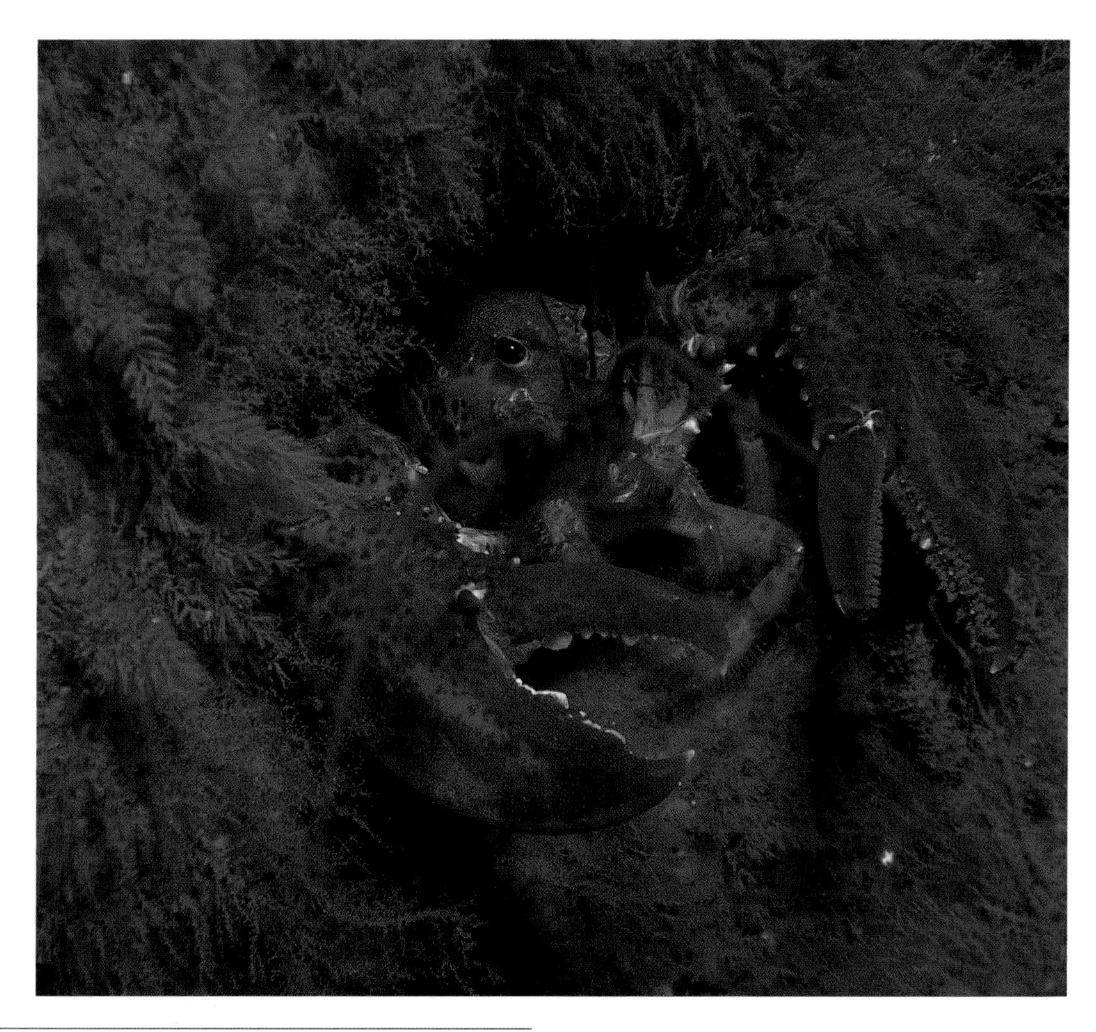

The North American lobster is perhaps the hallmark species of the New England shore. They use their large claws largely for defense and during jousting matches.

The North Atlantic coast of the United States begins at the border of Nova Scotia and Maine and ends at Cape Cod, Massachusetts. It includes the New England coast, as well as Nova Scotia, New Brunswick and Newfoundland outside of the U.S. The northern reaches of this cold-temperate region are exposed to the even colder Arctic region. At the southern end of the North Atlantic coast, Cape Cod acts as an effective barrier to marine life that might ride the Gulf Stream up from the southeast coast in the summer. It also keeps much of the cold Labrador Current from extending further south.

Rocks are a primary element of the seascape of the North Atlantic shore. In Maine, the rocks form a rampart against the sea and provide a foothold for the gnarled tree roots that seem to grasp at them. The rocky coast is one of sharp irregularity—islands, jutting headlands, deep bays, numerous peninsulas. If you followed every curve and twist of the shoreline, the distance between Portland, Maine, and the Canadian border would be 2,500 miles, whereas the straight-line distance is a mere 200 miles.

The myriad offshore islands on the Maine coast are the summits of hills drowned by the advance of the ocean upon the land. The islands are not arranged at random. In some places they appear as long, straight rows pointing toward the sea, moving into deeper and deeper water until finally they disappear under the waves. If you stand onshore and turn from the islands to look toward the hills, it becomes clear that both are actually the same range, some of which are partly covered by a rise in sea level.

The rugged northeast shore extends northward from Massachusetts with few interruptions. Desolate bare rock and gray sea suggest a throwback to an earlier chapter in Earth's history. Little vegetation lives on these granite headlands, but crowded seabird colonies dot the ledges. When John James Audubon visited the northeast shores in 1833 he found them so crowded that the nests nearly touched one another. Many kinds of birds are nearly as abundant today. Among the most attractive are the gannets, found from the St. Lawrence northward and best seen at the Avalon Peninsula of Newfoundland and at the Bonaventure Island off the Gaspé Peninsula. On jagged, precarious ledges shrouded in fog and drizzle, these magnificent birds raise their young. Their plumage is gleaming white, their backs tinged with gold, and they have a wingspread of six feet. From the ledges they flap out to sea, spending hours on the wing to reach feeding grounds that are sometimes more than a hundred miles away.

There is, for all their desolation, an air of newness about these rocky shores, for it was only recently in geologic time that the glaciers retreated and conifers reclaimed the land. Great stretches of the Labrador and Newfoundland shore are still barren of trees, and Newfoundland's fauna is still impoverished. Out of approximately 790 North American bird species, only 121 nest in that province. There were no moose until they were introduced by people half a century ago. Even the mallard duck, a hallmark species of the north, is absent.

The ruggedness of the rocky coast is due mostly to its geologic youth, for the granite rocks have been largely resistant to erosion. There are, however, some striking pillars, arches and sea caves. The erosion of the shore by waves is particularly noticeable in New Brunswick and northern Nova Scotia, where the waves break directly against less-resistant sedimentary

Rocky reefs are the most diverse habitats in the Northeast, providing a hard substrate where animals like these soft corals and sea anemones can attach, grow and provide shelter for a variety of smaller organisms.

rocks. Here, the shore is cut back at a rate of six to twelve inches a year, leaving the more resistant rocks as headlands thrusting into the sea.

One of the most notable places to see the effect of battering waves is the unsheltered island of Monhegan, Maine, whose steep cliffs lie directly in the path of storm waves. During heavy seas the crest of White Head, about 100 feet above sea level, is drenched with spray. Another wave-battered spot is Perce Rock at Gaspé—a long, narrow island that has been pierced by an opening through which small boats can sail. Here the cliffs, rising vertically 300 feet above the whitecaps, have been carved into the shape of a tall chimney that faces out to sea.

Mount Desert

Mount Desert is the largest of the islands that dot the Gulf of Maine. Here you'll find the highest point of land (1,532 feet above sea level) on the entire Atlantic seaboard north of Rio de Janeiro. Today's shoreline of Mount Desert is a result of the constant erosion of rock due to wave action, creating sea cliffs and undercutting blocks of granite from their foundations, resulting in one of America's most scenic coastlines.

Soft Shores

Traveling along the rocky northeastern coast, you'll discover small sand and gravel beaches, some formed from glacial debris, others built up by the action of storm waves. In sheltered places on the shore, waves and currents deposited huge quantities of sand, gravel and small boulders. These were added to the materials cut from the weaker rocks along the headlands, forming the narrow beaches that border the cliffs. Such cliffs and beaches are characteristic of a youthful shore (one that has not had time to accumulate significant amounts of sediment). As the shoreline matures, the waves continue to deposit material in the relative calm of coves and bays. At the same time, rivers fill in many of these places by bringing down particles of rock from land. At several places in northern New England, the jagged coast has been smoothed by the wearing back of the headland cliffs and the filling in of bays.

During this process, many of the islands lost their individuality and became tied to the mainland by sandbars. New Englanders usually refer to these sand bars as "necks." One of the best examples is at Marblehead, Massachusetts. At one time there was a large, rocky island offshore.

Goosefish are voracious predators, feeding on other fish, crabs, shrimp and even small seabirds!

WAQUOIT BAY

Habitats
Salt ponds
Salt marshes
Pine and oak forest
Barrier beaches/dunes
Open water

Key Species
Piping plover
Least tern
Sandplain gerardia
Alewife
Winter flounder
Blue crab
Osprey

Description
Waquoit Bay NERR includes: South Cape Beach, a 500-acre state park recreational area; Washburn Island, with 330 acres and 11 camp sites and the Sargent Estate headquarters. A 450-acre riverine pitch pine/scrub oak uplands tract protects the watershed. These areas are connected by open water, tributaries, salt ponds and about 15 miles of shoreline totaling 2,250 acres.

Location
Midway between towns of Falmouth and Mashpee, Massachusetts, on south shore of Cape Cod

WELLS

Habitats
Fields and forests
Tidal rivers
Swamp forest
Salt marshes
Dune forest
Beach

Key Species
Whitetail deer
Spartina grasses
Slender blue flag
Snowy egret
Soft shell clam
Winter flounder
Piping plover

Description
Wells features a saltwater farm with historic buildings and a 7-mile trail system winding through fields, forests, wetlands, salt marshes, dunes and beach. The Greek Revival-style house accommodates a visitor center and an estuarine exhibit.

Location
Route 1, Wells, Maine

But as waves and currents eroded it, the pebbles, sand and gravel that were chipped off added to the other sedimentary debris brought by currents to the island. Gradually a sandbar was built up that caught additional loads of debris. As it reached out toward the mainland, Marblehead Neck was created. Similarly, in the northern part of Boston Harbor two islands, Big and Little Nahant, were first tied together by a neck, and, later, still another neck attached them to the mainland.

Cape Cod

Cape Cod is the best known of the sand beaches that interrupt the rocky northeastern shore. Its bent arm thrusts far out into the ocean, presenting a formidable obstacle to the mixing of warm water from the south with cold northern currents. Cape Cod is also a barrier to the movement of marine life. Organisms that exist on the north and south shores of the Cape, sometimes only a few thousand feet apart, often represent quite different communities of marine life.

Cape Cod is an outstanding example of the sea's ability to create new landscapes. It owes its existence to raw materials provided by the erosion of the land by glaciers and to the force of tides and currents.

The irregularities that once existed in Cape Cod's shoreline have been smoothed out over time, and the famed beaches stretch for miles in nearly straight lines. They are now in a stage of destruction, each wave removing the sand and sediment forming the beaches. In some places the shore is receding by as much as three feet a year, although the average rate is much slower. Perhaps in another 5,000 years, the Cape will be a submerged bank. This process could be accelerated if global warming trends continue, raising sea level as the shoreline recedes through wave erosion.

This whelk is cruising along on its large, muscular "foot." The long tubular unit sticking straight out of its head is its siphon; the smaller elongated organ pointing down is one of its two "tentacles."

GREAT BAY

Habitats
Upland fields and forests
Salt marshes
Mudflats
Rocky intertidal
Eelgrass beds
Channel bottom and subtidal

Key Species
Bald eagle
Rainbow smelt
American oyster
Horseshoe crab
Large salt marsh aster
Osprey

Description
Great Bay is a "drowned river valley" estuary composed of tidal waters and mudflats. The reserve includes 1,375 acres of traditional New England salt marshes, mixed wood uplands and fields that have been logged and cleared since colonial times. The area includes a national wildlife refuge.

Location
10 miles inland from the coast of New Hampshire and the Maine border

North Atlantic Islands

The same geologic and physical forces that created Cape Cod molded many other places on the East Coast southward to New York City. Lying off the southern shore of the Cape are Martha's Vineyard, Nantucket and the Elizabeth Islands. The shape of most of these islands is roughly a triangle, with the long base at the south. Glacial events created the ridges that form the apex and sides of the triangle. Between the sides sandy outwash plains slope gently to the sea. Most of the islands have rough, jagged southern shores resulting from the rise of the sea after the melt of the glaciers.

Hermit crabs are some of the most personable members of the undersea world. This species is living in a moon snail shell covered with a colonial hydroid.

Long Island, New York, resembles a giant fish with a sweeping forked tail, its head pointed toward New York City. This fantastic shape has its origin in the glacial moraine that stretches along the hilly northern portion of Long Island to form the fish's backbone. All the land south of the moraine is a flat, sloping outwash plain. Toward the eastern end the moraine divides, one spur forming the upper fin of the tail, Orient Point, the other the lower fin, terminating with Montauk Point. The two peninsulas are the result of a minor retreat by the glacier. First the ice advanced as far south as Montauk, where it remained stationary for several thousand years, creating the southern spur of the moraine. Then the ice began to melt, only to become stabilized again a few miles to the north, producing the upper fin.

The moraine continues westward from Long Island to Staten Island and into New Jersey. It is breached by a channel called the Narrows, which forms the entrance from the Atlantic Ocean into the New York Harbor. If it had not been for the power of the Hudson River in breaching the moraine and keeping the Narrows open, New York Harbor and the metropolis that grew up around it would not exist.

The two rows of tiny black dots lining the edge of this scallop are its eyes, which alert this mollusc to approaching hazards like hungry skin divers.

From Staten Island the moraine follows a generally westward direction, marking the most southerly influence of the ice sheets upon the Atlantic coast. The contrast is striking as you pass from the steep, jagged sea cliffs of Maine to the narrow, boulder-strewn beaches of New England to the wide, sandy beaches of New Jersey. And with the change from rock to sand, from pounding surf to rhythmic breakers, there is a marked and fundamental difference in the assemblages of life that have adapted to these different shores.

The Importance of the Ocean Floor

The ocean floor is one of the most influential physical characteristics affecting the diversity and richness of marine life. It boils down to one of two types—hard or soft sea bottoms. The Maine coastline is dominated by a rocky shoreline, offering a hard surface to its marine flora and fauna. South of Cape Cod the coast is predominately sand, providing a soft surface to its local marine inhabitants.

Hard surfaces are made up of rocky headlands and reefs (in tropical regions animals such as corals may create their own hard surfaces, but that is another story). They also come in the form of artificial structures such as docks and pilings. Even sunken ships provide firm, though relatively temporary, hard surfaces. All of these provide a place for sea life to attach, like sea anemones, corals, tubeworms, hydroids, bryozoans, sponges and others. These plants and animals have a life cycle that at some point includes a stage when either adults (usually) or juveniles (sometimes) are firmly attached to a hard surface.

Animals fixed to the marine bottom have two major requirements. First, they must be able to disperse their young to colonize other suitable habitats. This is accomplished by water currents that transport microscopic larvae released by the attached adults. Second, the adult must be able to eat. Water also provides a source of nutrition for attached animals by bringing food to them on the same water currents that take away their progeny. Most attached marine animals feed on tiny, drifting plants and animals called plankton, trapping it either by filtering the water, like sponges, or actively trapping or catching passing plankton with mucus nets, filaments, tentacles and other devices.

These two American ocean pouts are doing what they do best—pouting around the seafloor waiting for an unsuspecting shrimp, crab or small fish to come within striking distance of their huge mouths.

In places where hard surfaces are available, such as along the Maine coast, competition for turf is fierce among marine plants and animals. Usually every square inch is covered with some kind of encrusting, attached "thing" that bears little resemblance to land animals. Touch is more important than sight to an attached animal. The sea anemone, for example, waits for its meals to bump into it rather than going out and running down its dinner.

The colonization of a rocky reef by attached animals and plants creates a complex, multi-structured environment that provides niches for all kinds of marine life. The hard tubes of a tubeworm colony provide yet another surface for the growth of a sponge. The soft sponge provides a home for a shrimp that can burrow into it. With the addition of each new colonizing species the habitat as a whole becomes more complex and the diversity of marine life increases. The attached animals and

Lumpfish are odd-looking piscines, which, in the absence of artificial light (like a photographer's flash), blend in with a background of rocks and seaweed.

plants not only provide food for passing fishes, which soon begin to congregate around the maturing community, but places of refuge for young fishes and small, errant organisms such as crabs and free-crawling worms. Eventually the mass of life attached to the rocky reef creates its own suite of "microhabitats" simply through the sheer biomass of life accumulating on it. For those reasons, hard surfaces provide the earth's richest and most diverse marine environments.

Filter-feeding animals dominate the subtidal attached communities. These are the animals that, like barnacles and mussels, feed on plankton. Sponges draw water into their bodies through tiny pores all over their surfaces. They trap the food particles contained in the water, then expel the filtered water through the large openings visible on the surfaces of many kinds of sponges. Sea squirts resemble sponges, having a pair of siphons leading into a chamber containing a basket that traps the food particles. Still other filter-feeders employ variously equipped arms or tentacles to catch plankton. These include the bryozoans, certain kinds of worms, sea cucumbers, brittle stars and brachiopods. Filter-feeding species of sea cucumbers in the Gulf of Maine include *Psolus fabricii,* a bright red animal, and the much larger, brown-colored *Cucumaria frondosa*. These animals—which belong in the phylum Echinodermata with starfish, sea urchins and brittle stars—fasten themselves to rocks with their tube feet and spread bushy tentacles to entangle the plankton that sweeps by in the tidal currents. They then stick each tentacle into a central mouth and suck off the trapped food particles.

Moon snails are conspicuous members of coldwater, sandy subtidal habitats on both the East and Westcoasts. They actively hunt and feed on clams as well as scavenge on any dead marine life they encounter.

Among the echinoderms, certain species of brittlestars are also filter feeders. The filter-feeding brittlestars tuck themselves away in crevices in the rocks and spread their five arms into the water. Comb-like processes that extend outward along the edges of the arms trap drifting plankton. The brittlestar then transports the particles of food to its mouth at the center of its button-like body.

Brachiopods, although they superficially resemble clams and other bivalves, differ in internal anatomy and belong to

Atlantic spadefish have one of the largest geographic ranges of any American coastal fishes. They can be found from Cape Cod to Brazil, as well as the Gulf of Mexico.

another phylum. Brachiopods attach to rocks by a stalk, and catch plankton with a tentacle-like organ known as a lophophore. Bivalves, on the other hand, feed by pumping water through their gills. The fossil record indicates that hundreds of millions of years ago the brachiopods constituted one of the dominant groups of animals in the ocean. Since then, however, they have steadily lost ground—in numbers of species as well as in absolute numbers—through competition with the more efficient bivalve mollusks.

Other rocky shore animals that feed with tentacles include hydroids (such as Tubularia and Clava) and sea anemones, like the ubiquitous Metridium. These are relatives of the jellyfish and members of the phylum Cnidaria. The tentacles of this group have tiny stinging capsules known as nematocysts scattered over their surfaces. Food particles coming into contact with the tentacles are penetrated by a hypodermic-like apparatus that pumps poison into the prey.

Rocky Reef Dwellers

Northern starfish are the most prominent, large invertebrate predators of the subtidal zone and move actively over the rocks in search of their favorite food, the blue mussel. To eat a mussel, the starfish humps up over its prey and grips it firmly with its tube feet. The starfish then sticks its membranous stomach through its mouth and manages to insert it into the narrow gap between the mussel's shells. As a final insult, the starfish digests the soft parts of the mussel within its own shell. When digestion is complete, the starfish pulls its stomach back into its body cavity, relaxes its grip, and the shells fall free.

The shells of the mussels are often used by a relative of the starfish, the green sea urchin, *Strongylocentrotus droebachiensis*, which attach mussel shells to themselves with their tube feet. It is possible that shells protect the urchins from the ultraviolet rays of the sun, or act as camouflage. The sea urchins feed by browsing on the algal film that covers the rocks. Where sea urchins are common, they play a major role in keeping the rocks free of attached algae, making room for other sessile organisms.

Rock crabs, *Cancer borealis,* and lobsters, *Homarus americanus,* lurk in crevices among the rocks and prey on mussels or scavenge dead animals on the bottom. Lobsters are very common in summer, when they occur in enormous numbers, even close to the low-tide mark. Along the shores of the Gulf of Maine, the fishing pressure on lobsters is so heavy that few grow much beyond the legal size limit. Here and there, however, isolated individuals survive and manage to grow larger.

Lobsters are caught as far south as Cape Hatteras, North Carolina, but

This slender starfish is crawling over a rock covered by pink coralline algae. The green sea urchins feed on green algae that also grows on rocks and other hard surfaces.

lobstering reaches peak density along the coasts of Maine, Nova Scotia and the southern Gulf of Saint Lawrence. They are caught in depths ranging from 1,200 feet to the intertidal zone. Lobsters, like all crustaceans, must molt or shed their hard shells in order to grow. Molting usually occurs in late spring or early summer, depending primarily on water temperature. The shell splits down the top midline and across the back where the tail joins the body. The lobster backs out of the old shell. Newly molted, the lobster's shell is exceptionally soft and wrinkled. Initially unable to move or defend itself, the lobster is at this time vulnerable to predators, including other lobsters. Within six hours the lobster has taken in enough water to completely expand the new, soft shell, which then begins to harden over the course of 7 to 10 days.

Lobsters mate when the sexually mature female has recently molted, but before her shell hardens. The female produces thousands of eggs, attaching them to the swimmerets on the underside of her abdomen or tail. There the developing eggs are protected by the heavy plating of the tail shell and constantly aerated by the fanning motion of the swimmerets. The female will carry these eggs for nearly a year until they hatch and become members of the plankton.

Depending upon temperature, the larval lobsters remain in the plankton for one or two months. During this time the young lobsters molt three times. After pthe third molt, most larvae abandon the plankton and gravitate toward the sea floor. However, it is not uncommon to find some fifth-stage larvae still in or about floating seaweed. They, too, eventually become bottom dwellers, leaving the plankton and joining life on the bottom.

Once on the bottom, the young lobsters' struggle for survival intensifies. Crabs, lobsters and several species of fish, including the longhorn sculpin, the sea raven and the winter flounder, feed on them. During their first season these young may molt as often as 12 times. Size at sexual maturity depends upon such variables as food and bottom temperatures. It takes lobsters between 6 and 8 years to reach lengths of 10 inches.

The Character of the North Atlantic Continental Shelf

The continental shelf is that part of the sea floor that begins at the shoreline and slopes down gradually to a depth of about 600 feet. At this depth the sea floor drops at a more dramatic angle, forming what is called the continental slope, which continues down to the abyssal sea floor. On the East Coast of the United States the continental shelf is a broad plateau, much of which was high and dry during the last ice age, when sea level was significantly lower than today. In some places the shelf drops quickly, such as off the southern coast of Maine, but this is the exception. In most places along the East Coast the shelf forms a broad, expansive undersea plain. Off New York and Massachusetts the shelf extends over 100 miles offshore before dropping off into the abyss. The same is true even farther south off the coasts of North and South Carolina and Georgia.

The relative shallowness of the shelf has a profound effect on the types of marine life found in the Northeast. Seasonal changes in temperature and weather patterns have the greatest effect on the marine life of this region. Here changes in air temperature and pressure regulate the sea's surface temperatures directly. The differences between the northeast winters and summers are felt just as dramatically under the water as above it. Water temperature off New York may rise to 60°F in the summer, plummeting to a near-freezing 38°F in the winter. Such a variation puts severe limits on the kinds of animals and plants that can live in such an environment. The result is a general decrease in the diversity of marine life, and a marked change in the kinds of organisms present from one season to the next.

Ocean Currents and the Distribution of Marine Life

Ocean currents also play a major role in the distribution of marine life along the northeastern seaboard. As the Florida Current passes Cape Hatteras on its northeasterly course it becomes the Gulf Stream. Previously, it had been flowing near or over the continental shelf, but beyond the cape it extends past the continental slope and into deep water. It is composed of a series of meandering filaments, some of which may rip along as fast as 5 knots. For most of its course the Gulf Stream stays well out to sea, but it approaches shallow water once more over the Grand Banks, which extend to the southeast of Newfoundland.

Zoogeographically speaking, the most important function of the Gulf Stream is to form an effective barrier between a cold-water area to the northwest and the warm water of the Sargasso Sea (the western North Atlantic) on the southeast. Between the Gulf Stream and the East Coast there lies a southwest-flowing, coastal current that tends to form an elongated, counter-clockwise eddy. The coastal current is supplied with cold water from the Labrador Current, which flows south out of the Labrador Sea and around the tip of Newfoundland. It is this coastal current that has in large part determined the nature of the marine life that occupies the shelf from Newfoundland Island south to Cape Hatteras.

Seasonality and the Distribution of Marine Life

Often called the "middle Atlantic seaboard" by marine zoogeographers, the region between Cape Hatteras and Cape Cod is penetrated during the summer months by a huge number of tropical and warm-temperate fishes and invertebrates that ride along on the warm, northeasterly flowing Gulf Stream. As fall and winter approach, these warm-water animals are, for the most part, left stranded to perish in the rapidly cooling waters of the north as the Gulf Stream's influence decreases and the weather patterns of the north begin to control the water temperatures of the shallow East Coast continental shelf. Because of the extreme seasonal fluctuation in temperature that takes place in the shallow waters of the area between Cape Cod and Cape Hatteras, it would be difficult for animals from either the north or south to become permanent residents of this dynamic coast.

Cape Hatteras represents the boundary between the warm-temperate and cold-temperate faunas. At this point on the coast the shelf narrows, extending out from shore less than 100 miles. This is also where the Gulf Stream turns northeast, heading out to sea. Evidence of this biological boundary exists most vividly in the distribution patterns of the most mobile marine animals—fishes. Research on Gulf of Maine fishes shows that of the

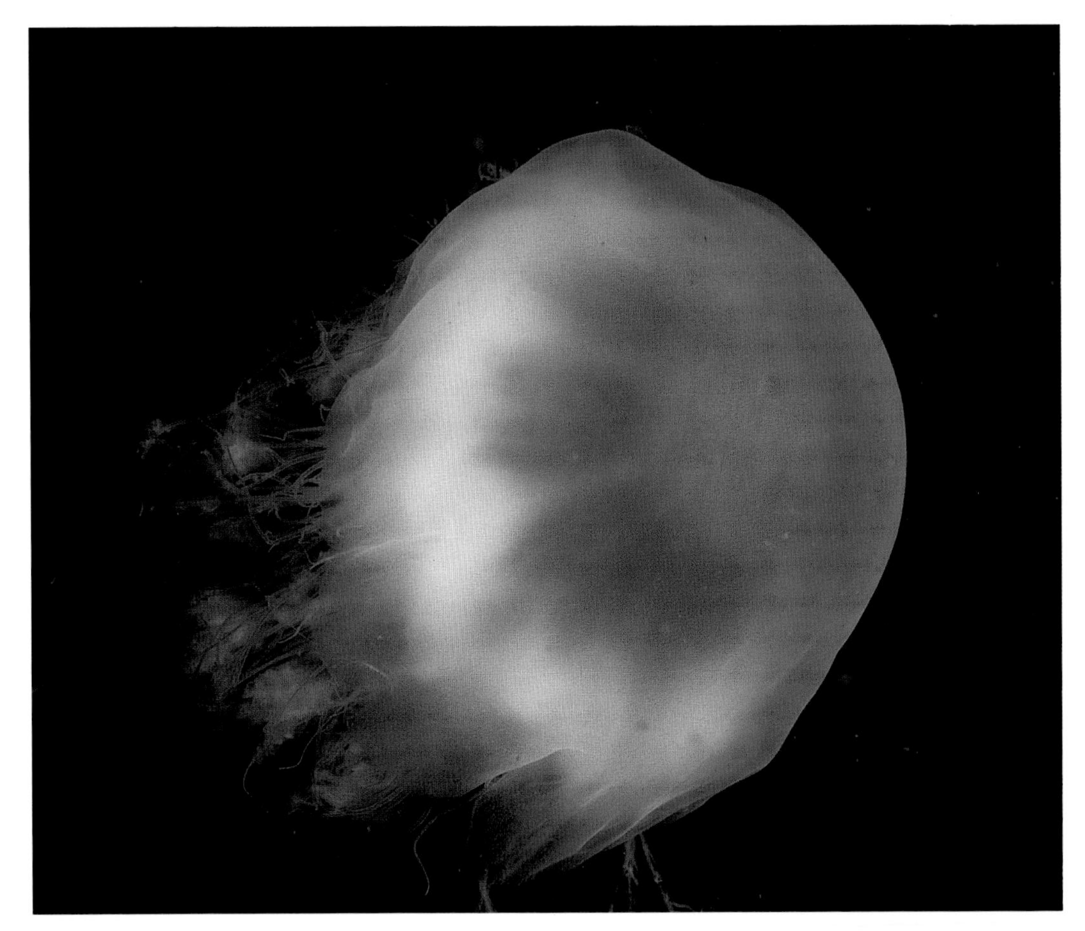

One of the world's largest jellies, lion's mane jellies have been found with bells three feet in diameter and tentacles fifty feet long.

129 species found along the shore, 20 percent are endemic to the area between Labrador and Cape Hatteras, 25 percent are shared with the Arctic and 55 percent are warm-water species of only occasional occurrence. In other words, only about 15 percent are found to live exclusively within the region north of Cape Hatteras, while nearly half of the fishes found there are transients originating in the south.

In midwinter, when water temperatures are at their lowest, few fishes are evident inshore except the odd sea raven, (*Hemitripterus americanus*). In March, when water temperatures rise to the neighborhood of 35°F a dramatic change occurs. Eelpout (*Macrozoacres americanus*) arrive in the shallows, indicated by little piles of sand dollar shells on the bottom that they have regurgitated. Later when the water has warmed sufficiently, eelpout seek cooler, deeper water. At about the time that eelpout first appear, flounders, (*Pseudopleuronectes americanus*) and little skate (*Raja erinacea*) make their appearance and will remain until the end of fall.

When water temperatures rise to the neighborhood of 47°F, the cunner (*Tautoglabrus adspersus*) turns up out of nowhere. Alewives (*Pomolobus pseudoharengus*), about a foot in length, congregate at the mouths of freshwater streams prior to ascending them to spawn. At about this time lumpfish (*Cyclopterus lumpus*) also collect inshore. They usually attach to rocks by their pelvic fins, which are specially modified to form suckers.

As spring turns to summer, huge numbers of spiny dogfish (*Squalus acanthias*) work northward into the shallow waters of the gulf. Around midsummer small mackerel (*Scomber scombrus*) and pollock (*Pollachius virens*) school close to shore. Squirrel hake (*Urophycis chuss*) move over the bottom, sensing every inch of the way with pelvic fins strangely modified for use

STELLWAGEN BANK

Habitats
Sand and gravel bank
Muddy basins
Boulder fields
Rocky ledges

Key Species
Northern right whale
Humpback whale
White-sided dolphin
Storm petrel
Northern gannet
Bluefin tuna
Atlantic cod
Winter flounder
Sea scallop
Northern lobster

Description
Formed by the retreat of glaciers from the last ice age, Stellwagen Bank consists primarily of coarse sand and gravel. Its position at the mouth of Massachusetts Bay forces an upwelling of nutrient-rich water from the Gulf of Maine over the bank—leading to high productivity and a multi-layered food web with species ranging from single-celled phytoplankton to the great whales.

Cultural Resources
1898 wreck of the steamer *Portland*
Middle Ground fishing area

Location
25 miles east of Boston, 3 miles southeast of Cape Ann and 3 miles north of Provincetown

Protected Area
842 square miles

as organs of taste and touch. It is at this time that the first goosefish (*Lophius americanus*) come into the shallows. They lie flush with the bottom, partially covered with sand. Little lobes of flesh all around the edges of the body help them blend in beautifully with their surroundings. Only their red-green eyes and huge, grinning, upturned mouths give them away. The goosefish has a leaf-like flap of skin on its first dorsal fin ray, which it waves back and forth over its mouth to entice other fish near enough for it to strike. Goosefish are ambush predators, and they are not very good swimmers. Despite their ungainliness, they may swim to the surface, where they have been known to engulf birds in their enormous jaws.

In late summer, the alewives that spawned the preceding spring in ponds at the headwaters of streams return to the sea. When they do, they are about three or four inches long and move in the surface waters, snapping at individual particles of plankton.

Invertebrates also wax and wane seasonally. In midwinter and spring, *ctenophores*, such as *Mnemiopsis* and *Pleurobrachia,* maintain their position in the water column by beating eight rows of shimmering combplates. Two-inch-long annelid worms writhe in the surface layers. Caprellid amphipods swim up toward the surface with humping motions, then drift downward with outstretched pincers or claws to snatch their prey. Isopods gyrate crazily through the water and the pteropod, or winged snail (*Clione limacina*), paddles gracefully through clouds of darting copepods in the upper layers. Later in the year the moon jelly (*Aurelia aurita*) appears in enormous numbers. Scattered in the jellies' midst are small Arctic red jellyfish, or lion's mane (*Cyanea capillata*). Later still and far offshore, *Cyanea* grow huge, with bells three to six feet in diameter.

With the onset of fall and cooling of the waters, there is a progressive decrease in the abundance of fish and invertebrates. Finally only the occasional sculpin (*Myoxocephalus octodecimspinosus*), or sea raven, is to be seen. By this time storms have torn loose and swept away the seaweed, and the waters of the northeast coast chill through the winter until the beginning of spring, which will initiate a fresh cycle of growth and abundance.

A truly bottom-dwelling fish, sea robins literally crawl about sandy bottoms on special pelvic fins searching for food and mates. They range along the entire east coast from the Bay of Fundy to Palm Beach, Florida.

TWO

The South Atlantic Coast

Miles of sandy beaches characterize the shores of the southeastern seaboard, stretching from the Chesapeake Bay to the Florida Keys. The sand, composed of mineral particles that range in diameter from about two-thousandths to eighty-thou'sandths of an inch, comes from the erosion of rocks containing limestone, feldspar, quartz and other minerals. The sandy shoreline on this part of the coast is due primarily to the presence of large river systems. The Potomac, for example, has deposited sediment originating from inland geologic processes on the coast for centuries.

A typical sand grain may arrive at the beach from a distance of several hundred miles inland, where perhaps a boulder has been torn loose from a mountainside. As the boulder is fractured and splintered, many of the resulting particles are deposited in rivers and carried to the sea. The sand grains that complete the trip to the beach may make it in a few days, or they may travel for several years. Wave action and currents do the final sorting according to the weight and size of the individual particles. The lighter material is carried along the shore by waves and currents and deposited on another shore. The heavier particles are flung back upon the upper beach or swept out to deeper water by strong ocean currents; there, they gradually move down the continental slope into the abyss.

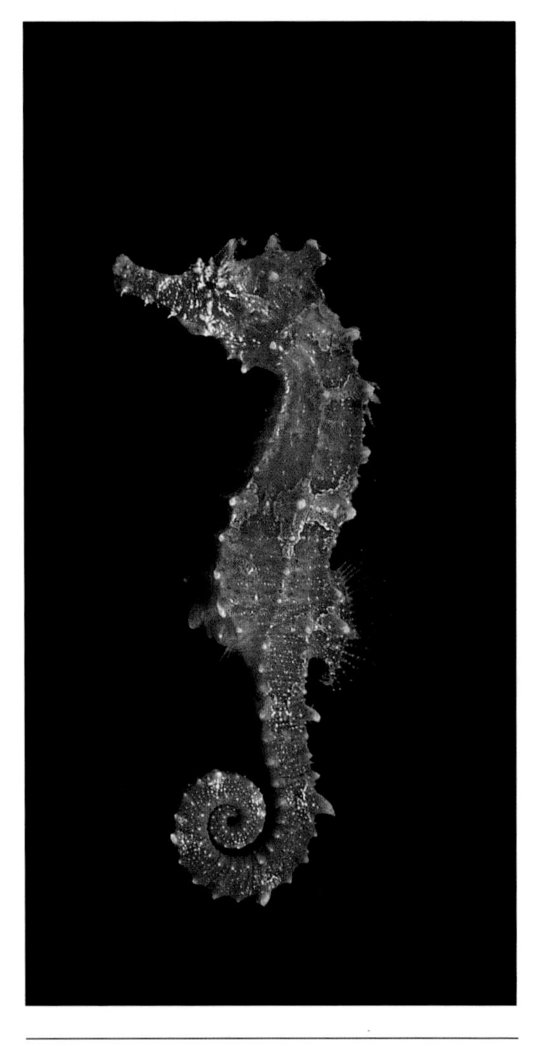

Icons of the ocean realm, seahorses don't really seem to be fish at all but rather a collection of body parts from a variety of unrelated animals. This species, Hippocamus hudsonius, *can be found from South Carolina to Cape Cod.*

The composition of sand differs geographically according to the type of rock from which it was derived. The sand beaches from New Jersey northward owe their origin largely to a mixture of minerals carried shoreward by glaciers. Further south, the proportion of limestone in the sand increases. Along the North Carolina coast as much as a tenth of the sand may be composed of particles of broken shells. Southward from North Carolina the sands again are of mineral origin, derived from the weathering of rocks on the eastern slope of the Appalachians. On a number of Florida beaches, the ocean currents have tossed up large quantities of quartz. These beaches are famed for their glistening whiteness and uniformity, and also for their hardness, as exemplified by the packed sands of Daytona Beach, which are virtually pure quartz. The composition of the beach sand changes once again in southern Florida, where quartz becomes mingled with shell and coral fragments. On the Gulf Coast the sands usually contain a large proportion of mud and decaying plant material.

These vast stretches of sand seem like an inhospitable setting for marine life. In contrast to a rocky coast, the sandy beach uncovered at low tide reveals a seemingly barren world. Instead of the crowded life that fills the tidepools and crevices of a rocky shore, the sandy beach at low tide reveals only the marks of the waves and the flotsam and jetsam that accumulate on it. The shifting sands of the beach afford no surface to which either marine plants or marine animals may attach and gain a foothold. Nowhere is there any change in relief from a flat, homogenous topography. All life forms that exist in soft surfaces, such as mudflats and sandy beaches, must have adaptations for digging and burying themselves in the bottom itself in order to escape both detection from predators and the deadly effects of drying when the beach is exposed at low tide. All the marine life on a soft shore is living in the beach, not on it, and it will take a little digging to discover it.

If you look closely, you can see subtle evidence of the abundance of life within the sand. On coasts with little wave action, the beach may be peppered with the tiny holes of ghost shrimp. The trails you may see on the sand are not caused by wind or tide but by a burrowing animal, like a predaceous worm searching for another worm that will become its prey. Besides being able to dig to bury itself, burrowing animals also need ways to breath and feed, either above or below the surface. Some, such as clams and ghost shrimp, develop tubes and burrows that connect them with life above. Others remain completely buried. These extract oxygen from the water that seeps through the sand and feed directly on the sand itself, just like an earthworm does on land.

Life is further complicated for sand- and mud-loving animals by the fact that more often than not such environments are close to estuaries—places along the coast where the fresh water of rivers meets with the salt water of the ocean. By their nature, these are also shallow-water areas, and so fluctuations in both temperature and salinity can be extreme, even on a daily basis, with the exchange and mixing of waters with the tides. Perhaps the best example of such a system is the Chesapeake Bay, the largest shallow-water embayment on the East Coast.

Chesapeake Bay

An estuary is a semi-enclosed body of ocean water that is measurably diluted by freshwater runoff from the land. There are three main types; the fjord, such as Puget Sound in Washington; the bar-built estuary, formed by accumulating sediments, examples of which occur in the Gulf of Mexico and the coastal plain estuary, which is a drowned river valley. The Chesapeake is a coastal plain estuary with a surface area of over 3,000 square miles. Due to its shallow depths (less than 10 feet in most places) the physical environment of the Chesapeake estuary changes on a seasonal and even daily basis. At high tide salty ocean water rushes in, creating a wedge of less dense fresh water that pours over it. At this time the salinity of the estuary rises. At low tide the freshwater system has the upper hand and salinity decreases. Concurrently, there may be great changes in the temperature of the water as vast expanses of the estuary's soft bottom are exposed to air. In the winter this may result in some parts of the estuary freezing; in summer surface temperature may shoot up to 80° or 90°F. Clearly, organisms that inhabit estuarine waters must be capable of dealing with dramatic changes in the physical environment.

GRAY'S REEF

Habitats
Calcareous sandstone
Sand bottom communities
Tropical/temperate reef

Key Species
Northern right whale
Loggerhead sea turtle
Grouper/black sea bass
Angelfish
Barrel sponge
Ivory bush coral
Sea whips

Description
Just off the coast of Georgia, in waters 20-meters deep, lies one of the largest nearshore sandstone reefs in the southeastern United States. The area earned sanctuary designation in 1981, and was recognized as an international Biosphere Reserve by UNESCO in 1986. Gray's Reef consists of sandstone outcroppings and ledges up to three meters in height, with sandy, flat-bottomed troughs between. Because of the diversity of marine life, Gray's Reef is one of the most popular sport fishing and diving destinations along the Georgia coast.

Cultural Resources
Potential paleoindian artifacts

Location
17 miles east of Sapelo Island, Georgia

Protected Area
23 square miles

The American Oyster

The American oyster (*Crassostrea virginica)* is a mollusc that first appeared on the planet 200 million years ago and has changed little since. It has completely adapted to estuarine conditions, and as a result is enormously successful in this harsh environment. No discussion of the southeastern seaboard and the Chesapeake would be complete without describing this most hardy and delectable mollusc.

Not the most prepossessing of marine delicacies, this slimy creature led Jonathan Swift to remark, "He was a bold man that first ate an oyster." Because of its commercial importance the oyster has been the subject of remarkably extensive research. The American oyster is a bivalve found throughout much of Chesapeake Bay and similar estuaries, principally in water depths of 10–30 feet. The sex life of the oyster is a cyclical affair. During the winter its body weight is made up of the glycogen and salts that make it fat and tasty. But in the late spring, the oyster's physiology turns to reproduction and individual oysters become either male or female for the season, or at least part of it. The oyster at this time converts 80 percent of its body weight to sex organs, which are thin and watery and not particularly tasty. It isn't that oysters are inedible during the summer, it's just that they aren't as meaty and succulent. In the fall their body tissues reconvert to glycogen and salts, becoming asexual once more.

Sexually mature adult oysters release eggs and sperm into the water by the trillions throughout the summer. A single oyster may produce up to 100,000 eggs on a single spawn and may spawn four or five times a season. The following year this same oyster may turn male and emit millions of sperms.

Fertilization occurs by a chance meeting of a sperm and egg floating in the estu-

MONITOR

Habitats
Pelagic, open ocean
Artificial reef

Key Species
Amberjack
Black sea bass
Red barbier
Scad
Corals
Sea anemones
Dolphin
Sand tiger shark
Sea urchins

Description
On January 30, 1975, the nation designated its first national marine sanctuary. The site was the wreck of the *USS Monitor,* a Civil War vessel that lies off the coast of North Carolina. The *Monitor* was the prototype for a class of U.S. Civil War ironclad, turreted warships that significantly altered both naval technology and marine architecture in the nineteenth century. Designed by Swedish engineer John Ericsson, the vessel contained all of the emerging innovations that revolutionized warfare at sea. The *Monitor* was constructed in a mere 110 days.

Cultural Resources
The remains of the Civil War ironclad *USS Monitor*

Location
16 miles southeast of Cape Hatteras, North Carolina

Protected Area
1 square mile

ary. Fewer than 1 in 10,000 will survive the larval stage, thus the need to release such huge numbers of sex cells. Within a few days of fertilization the eggs develop into larvae. Each larva has a bivalve shell and a soft, muscular foot covered with hair-like ports that propel it through the water. For the most part, however, its movement and distribution within (or out of) the estuary is determined by the tides. Near the end of the second week, each larva has grown so large, its shell so heavy, that it becomes increasingly difficult for it to stay afloat and propel its mass through the water. The larvae drop to the bottom in search of a firm place to set for the rest

Wrapped in a crustacean embrace, these Jonah crabs are in the initial stages of mating. The male will carry his "date" around the ocean until the time is right.

CHESAPEAKE BAY, VIRGINIA

Habitats
Tidal waters
Marshes and flats
Submerged aquatic vegetation beds
Swamps

Key Species
Migratory neotropical birds
Sea grasses
All stages of life of commercially important fishes and shellfish and of their food

Description
At present, the Chesapeake Bay-Virginia Reserve features four components, all within the York River Basin. From the York's mouth moving upstream, they are the Goodwin Islands, Catlett Islands, Taskinas Creek and Sweet Hall Marsh. They represent the salinity gradient of the river and the range of habitats that result, including tidal saltwater and freshwater marshes, submerged aquatic vegetation, upland forests, beaches and open water.

Location
Within the York River basin

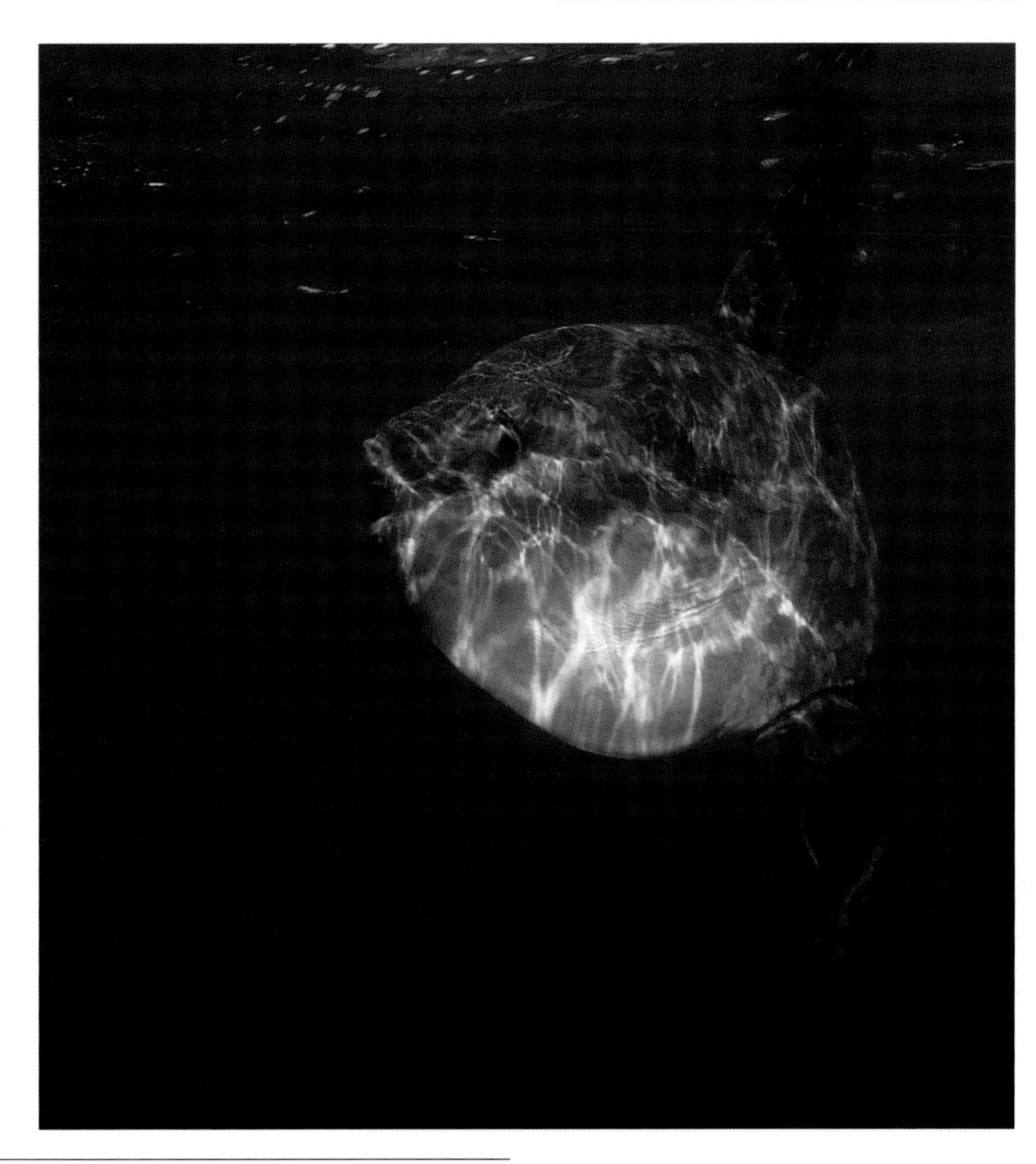

The ocean sunfish (also know by its scientific name as Mola mola*) is one of the world's most peculiar fish. Found in both the Atlantic and Pacific oceans, they feed mainly on jellyfish!*

NORTH INLET / WINYAH BAY

Habitats
Salt marshes
Intertidal oyster reefs
Mud flats and sand bars
Tidal creeks
Shallow sounds
Abandoned rice fields and canals

Key Species
Cordgrass
Fiddler crab
American oyster
Spot
Alligator
White ibis

Description
North Inlet / Winyah Bay features high-quality, ocean-dominated waters and salt marshes in North Inlet contrasting with brackish waters and marshes in the Winyah Bay. The bay's estuary is dominated by riverine discharges from a watershed impacted by agricultural, municipal and industrial development. Former rice fields and canals provide another system for study within the reserve.

Location
1 mile north of Georgetown, South Carolina

> **ACE BASIN**
>
> **Habitats**
> Salt marshes
> Brackish marshes
> Tidal flats
> Maritime forests
> Bird keys and banks
> Pine-mixed hardwoods
>
> **Key Species**
> American alligator
> Atlantic loggerhead turtle
> Shortnose sturgeon
> Bald eagle
> Wood stork
>
> **Description**
> ACE Basin features outer coastal plain communities typically associated with barrier islands, marsh islands and estuarine rivers. It is especially rich in its diversity of endangered and threatened species.
>
> **Location**
> 45 miles south of Charleston, South Carolina

of their 10–20 year lifespan.

This is perhaps the most critical point in the young oyster's life, for without a firm, hard substrate it will eventually sink into the soft bottom of the estuary as its shell grows and gets heavier. The problem is compounded by the fact that there are very few hard substrates in estuaries. In fact, the most abundant hard surfaces are the shells of other adult oysters, and this is exactly where most young oysters wind up—attached to their adult brethren. Remarkably, the senses of the young oyster, particularly taste and touch, are attuned to differentiating between the shells of other oysters and other kinds of hard substrates.

Once the shell of another oyster (dead or alive) is located, the larval oyster secretes a small dab of glue from its foot, turns on its side, presses its left shell valve into the glue and within 15 minutes the glue has set, and so, too, has the oyster—for life. Its foot, having served its purpose, is soon absorbed and the oyster begins a new and quite different life. No longer a member of the plankton, it has now joined the group of organisms that live on or in the bottom of the estuary.

Over time, and several generations of oysters, the accumulations of dead and living oyster shells form a significant mass that creates its own marine environment by providing a hard, complex surface where there would otherwise be a flat, soft bottom. These "oyster reefs" provide a place of attachment not only for oysters, but other forms of estuarine life as well, such as algae, tubeworms, sponges and sea anemones. The cracks and crevices of the piled-up shells also provide microhabitats for crawling organisms such as small snails, shrimps and crabs. Fishes are attracted to these reefs, feeding on the organisms that live there. In addition, young fishes use the reefs for cover as well as feeding grounds.

The newly settled oysters are called spat. They begin filter-feeding immediately, now eating the plankton from whence they came trapping it on mucus covering their gills. Over a 24-hour period an adult oyster may filter as much as 100 gallons of water, removing particles as small as a few thousandths of an inch in size. Within a month it will have grown to the size of a pea; within three months, to three inches. Within a year or two at most, the young oyster will be sexually mature.

Oysters thrive in water with salinities ranging from as low as 8 or 10 percent sea water to 35 percent. They grow more rapidly in the upper third of this range, but so do their worst predators, the oyster drill (a snail), starfish, boring sponges and cownose rays. Because the oyster's enemies are not able to tolerate lower salinity, oyster populations often concentrate in areas of relatively low salinity, those between 10 and 20 percent. The oysters may grow more slowly there, but they have a better chance of surviving in this habitat.

The Blue Crab

Like the oyster, the blue crab *(Callinectes sapidus)* is a hallmark species of Chesapeake Bay and the northern portion of the southeast coast. It, too, is primarily an estuarine animal, but unlike the oyster, it is highly mobile, and thus capable of moving about the estuary in response to temperature and salinity fluctuations in this ever-changing environment.

Blue crabs belong to the swimming crab family Portunidae. The most distinctive characteristic of this group is the enlarged and flattened hind legs that act as paddles to propel the crab sideways at a remarkably quick rate. Their claws are adapted for tearing and grinding apart any form of organic material they can find. Blue crabs are scavengers, and they contribute

NARRAGANSETT BAY

Habitats
Open water
Eelgrass meadow
Tidal flats
Salt marshes
Freshwater wetlands and ponds
Upland fields
Upland forests

Key Species
Bluefish
Striped bass
Winter flounder
Osprey
Great blue heron
Whitetail deer

Description
The Narragansett Bay Reserve includes undisturbed salt marshes, tidal flats, rock shores, open waters, upland fields, forests and a historic farm site. The reserve contains the major watershed and largest stream on Prudence Island. A deep water pier and recreational facilities are located at south Prudence.

Location
Approximately 10 miles south of Providence, Rhode Island

NORTH CAROLINA

Habitats
Salt/brackish/freshwater marshes
Sandy beaches
Shrub thickets
Maritime forests
Mud/sand flats
Subtidal vegetation
Oyster bars

Key Species
Cordgrass
Hard clam
Blue crab
Flounder
Loggerhead sea turtle

Description
The North Carolina reserve includes: Currituck Banks, a freshwater sound/marsh area containing ocean beach, dunes and maritime forest; Rachel Carson, with salt marshes, extensive mud flats, eelgrass beds and dredge material areas; Masonboro Island, a pristine 9-mile barrier island/sound complex and Zeke's Island, a man-made rock jetty with two distinct local estuarine habitats.

Location
Currituck Banks is 1 mile north of Corolla, North Carolina; Rachel Carson is across Beaufort, North Carolina; Masonboro Island is 1 mile south of Wrightsville Beach, North Carolina; Zeke's Island is 5 miles southwest of Kure Beach, North Carolina

significantly to the recycling of organic material in the estuary by breaking up large pieces of decaying animals and plants, thereby facilitating the work of bacteria which attack and complete the process of returning nutrients into the water.

Late summer is mating time for the blue crab. Sexually mature males are called "jimmies," females are "sooks." Males and females copulate, but impregnation is impossible until the female has shed her shell. Males seem able to detect when a female is going to shed and, not to miss an opportunity, latch onto the generally smaller sooks and carry them about the sea floor, cradled beneath their bodies until molting takes place.

Once the female sheds her shell she is in the soft-shell state, and impregnation takes place over a six-to-twelve-hour period. The male continues to cradle the female until her shell has once again hardened. At this point the romance is over, at least for the female, for this will be her one and only chance to mate this season. The males, on the other hand, mate repeatedly with other females.

With the onset of winter the crabs disappear from the waters of the bay. They slide off the shoals and down into the deep trough that marks the ancestral Susquehanna River Valley. The colder the winter, the deeper they go. For the jimmies, the water off Smith Point near the mouth of the Potomac, where the depths reach 100 feet or more, is a favorite spot.

However, wintering jimmies line the entire middle reaches of the bay extending as far north as Annapolis. The crabs back into the soft sediment at a 45-degree angle, and work their way in until only their antennae and eye stalks are left exposed. There they lie dormant, waiting for spring and sleeping through the entire winter.

The sleep of crabs who have been

With a face only a mother could love, this summer flounder was found partially buried in sand, its favorite habitat. Like all "flatfish," flounders actually lay on their sides, with one eye having migrated to the other side of the head.

fortunate enough to come to rest in Maryland waters, mainly males, will be undisturbed. Maryland law forbids dredging for crabs. But those bedded down in Virginia waters, mainly females, cannot rest easy. Virginia permits winter dredging for crabs and, for the eastern seaboard, the Virginia catch alone makes up 60 percent of the entire winter harvest. Many sooks will be rudely awakened, wrenched from their soft sea beds by a fisherman's dredge.

In May, the crabs emerge from the bottom, shake off their sediment-covered bodies and resume active life in the water. The business of the sooks is then, perhaps, the most crucial for the continued well-being of the species. The females open their abdomen and attach the eggs that were fertilized in the fall to their telsons, where they form a spongy orange-yellow mass. By mid-June the egg mass has turned from bright orange to dark brown or black, indicating that the eggs are ready to hatch. With several strong, deliberate snaps of the abdomen, millions of blue crab larvae are released into the water to become part of the plankton, at least temporarily. During the first six weeks of life, the first larval stage, called a zoea, molts several times. Each molt increases its structural complexity. At its final shedding the zoea takes the second larval form, a megalops, which looks like a misshapen lobster. The megalops is 0.06 to 0.12 inches long and is complete with claws and a tail. When it becomes a crab the tail will turn under against the abdomen to form the telson. As the megalops matures it begins its trek up the bay toward lower salinity, molting and changing as it goes until it becomes a tiny but recognizable blue crab. Like the oyster, the planktonic stages of the blue crab's life are the most hazardous.

In the first year the immature crabs will molt several more times, growing significantly with each shedding of their outer exoskeleton. By the summer or fall of the year following its appearance as a larval crab it is sexually mature, ready to make his or her contribution to the life cycle that sustains the species. Near the end of the summer many of the old sooks who have spawned leave the bay for the ocean to die. A few will live to return to the bay during the following summer, but they will never again mate. They are easy to spot. Their bodies are studded with barnacles and their once brilliant colors are dulled by sea moss. Three years is the maximum life span of a blue crab, male or female. Much of that time is spent sleeping; most of the rest is spent in search of food and a small portion looking for a mate.

Barrier Beaches

Most of the southeastern Atlantic and Gulf coasts are protected by strings of narrow, unstable barrier islands lying just off-

HUDSON RIVER

Habitats
Open water
Subtidal meadows
Tidal flats
Tidal marsh
Tidal freshwater wetlands
Mixed forest

Key Species
Grass shrimp
Herring
Blue crab
Snapping turtle
Muskrat
Osprey
Bald eagle

Description
The Hudson River reserve's network of four sites spans a 100-mile salinity gradient. Nearly 5,000 acres of tidal wetlands and uplands occur at Piermont Marsh, Iona Island, Tivoli Bays and Stockport Flats.

Location
Columbia, Dutches, and Rockland counties, New York

shore. Cape Kennedy is one of these barriers, which lies twenty miles from the mainland. Padre Island, off the southeastern coast of Texas, is about 140 miles long. The water between Padre Island and the mainland is a complex of ponds, islands and mudflats, providing a diverse habitat for a startling variety of shorebirds and typifying the estuary/marsh/lagoon environments created by these long, narrow sand islands. Over time these shallow-water habitats fill in with sediment, joining bar with bar and the bars with the mainland. Winter storms move large amounts of sand from the seaward side of the islands to the lee, or mainland, side. On the most narrow and low-lying islands storm waves may over-top the barrier and hurl sand behind it.

Running parallel to the mainland, these sandy beach islands form sheltered lagoons that form the Intracoastal Waterway, offering protected sailing almost without interruption from New Jersey to Mexico. One continuous lagoon, known as the Indian River, is about 130 miles long. The lagoons are filled in from the land side as river runoff carries sediment into the lagoon. Where the accumulation of sediment is greater than its transport out of the lagoon the barrier island and mainland will eventually become one. The lagoon disappears, replaced first by temporary swamps and marshland and eventually by dry land.

Familiar beach resorts such as Miami, Hilton Head, Daytona and Hatteras have been developed on barrier islands. Cape Hatteras is formed by a series of barrier islands, extending as much as 25 miles from the mainland, that shelter Pamlico and Albermarle sounds. The island is so narrow that at some places only a few hundred yards of sand lie between Pamlico Sound and the Atlantic Ocean. Cape Hatteras is the meeting place of the cold ocean currents that pour down from the north and the warm, northward-flowing Gulf Stream, which for the most part remains offshore. The meeting of these two water systems creates a substantial barrier to many marine organisms, whose annual distribution varies greatly depending on the dynamic interaction of these two water masses.

Sapelo Island

Perhaps one of the best examples of a classic barrier island is Sapelo, located off the coast of Georgia. In 1976, Sapelo was designated a National Estuarine Research Reserve in recognition of its natural and historical resources. Measuring only about ten miles in length and four across, Sapelo is separated from the mainland by a six-mile-wide belt of saltmarshes and tidal rivers. Sapelo Sound to the north of the island and Doboy Sound to the south connect the marsh rivers to the Atlantic. The ocean waves roll up on undisturbed, gently sloping sand beaches and mudflat's to the east.

Although most of Sapelo lies less than 10 feet above sea level, occasionally the land reaches 26 feet above the water. The terrestrial flora consists of oak and pine forests, interrupted with numerous freshwater ponds and swamps. Sapelo, like most barrier islands, probably originated from sand dunes and ridges formed landward of the shoreline by sediments deposited during glacial times, when sea level was lower and the shoreline was exposed. In the past 10,000 years the melting of the continental glaciers flooded the coast and submerged parts of the mainland, leaving the sediment ridges as barrier islands and creating lagoons landward of the barriers. Once the barrier islands have formed they may migrate or remain stationary, depending on physical conditions such as water currents and seasonal storm activity, particularly as it affects storm waves. The ocean may break through the barrier at points where the island is particularly narrow and unstable, creating new sounds that are the entrances to the lagoons between the island and the mainland. The width of the lagoon depends on the slope of the mainland, the degree of submergence and the balance between erosion and sedimentation from both land and sea. The lagoon is maintained only if submergence and sedimentation are slow and the water currents running through the lagoon (usually tidally driven) are sufficient to carry away the sediment that accumulates.

On the protected versus the exposed shores of Sapelo, the environments are completely different. The barrier lagoon consists of two parts, the estuary and the marsh. The estuary is a series of interconnecting, winding channels and saltwater rivers, usually only a few feet deep. The

SAPELO ISLAND

Habitats
Salt marsh
Maritime forest
Dunes and beaches
Freshwater ponds
Sloughs

Key Species
Cordgrass
Live oak
Osprey
Great blue heron
Wild turkey
Bald eagle
Loggerhead turtle

Description
Sapelo Island is the fourth largest Georgia barrier island and one of the most pristine. The reserve is made up of salt marsh, maritime forest and beach and dune areas. Not only is the island rich in natural history, but also in human history dating back 4,000 years.

Location
7.5 miles northeast of Darien, Georgia

sound connects them with the ocean. During high tide the water level rises, the channels become wider and deeper, and the water leaves its low tide channel and floods onto the marsh. As the tide ebbs thousands of creeks and tributaries drain the marsh, which returns to its low-water state.

The Sapelo estuary, as in all similar environments, is characterized by greatly changing physical conditions. The tide range measures 6.5 feet at Sapelo, but it exceeds 10 feet during spring tides. Full-strength seawater enters the lagoon through the sounds. The Altamaha River drains a large portion of the Georgia coastal plain and constantly carries its load of fresh water into the estuary five miles to the south of Sapelo's lighthouse. The daily tidal currents mix the ocean and river waters thoroughly. These same tidal fluctuations also set up swift currents in the narrow rivers and creeks of the estuary and marsh, attaining velocities of two knots or more on an average tidal exchange.

The marsh between the barrier islands and the mainland is a strange environment—neither marine nor terrestrial—at times baked by the sun, then flooded by the tide. While most terrestrial habitats are usually comprised of a variety of plants, the saltmarsh is completely dominated by one cordgrass species, *Spartina alterniflora.* In its living state, *Spartina* does not contribute significantly to the food web of the marsh community. Less than one-tenth of its biomass is used as food energy by other organisms while it is alive. Most of its contribution to the energy cycle comes after *Spartina* has died, falling into a tidal creek and decomposing. At this point microorganisms attack the leaves and begin breaking them down into smaller bits of particulate matter called detritus. With the next outgoing tide the decomposing grass and the microorganisms working on it are washed out of the saltmarsh and out to sea, where they contribute important nutrients to the offshore marine ecosystem. Thus the richness of the waters of the continental shelf depends to a large degree on the coastal marshes.

This link between the marsh and open ocean is also illustrated through the life cycle of one of Georgia's commercially harvested marine animals, the white shrimp, or *Penaeus setiferous.* White shrimp spawn offshore from April to September. When the eggs have hatched, and the shrimp have reached a length of two-fifths to three-fifths of an inch, they move inshore, where they remain in the sounds, tidal creeks and tributaries of the barrier island lagoon some three months or more until they are 4 to 5 inches long. In the relative safety of the estuary/marsh/lagoon system the juvenile shrimp feed and grow, eating mainly tiny crustaceans and detritus. Then in late summer the adolescent shrimp leave this marine nursery for offshore

CHESAPEAKE BAY, MARYLAND

Habitats
Forested uplands
Tidal freshwater marshes
Tidal brackish water marshes
Open waters
Coastal grasslands

Key Species
Peregrine falcon
Bald eagle
White perch
Sora rail
Baltic rush
Anglepod
Wild rice

Description
One site is not sufficient to represent the environmental diversity in the Chesapeake Bay. The three different components of the Chesapeake Bay, Maryland, reserve feature very different and distinct habitat types: freshwater and flooded hardwood marshes, brackish marshes, riverine wetlands and open water.

Location
All components are located within 200 miles east of Washington, DC.

waters, moving through the sounds and shallow coastal waters that connect the lagoons with the open ocean.

The marine environment on the seaward side of Sapelo is vastly different from that of the lagoon. Here temperature,

Horseshoe crabs aren't true crabs but ancient relatives that are actually more closely related to spiders and scorpions. In the summer they can be found by the thousands in Delaware Bay in a mating and egg-laying frenzy.

Unlike most sea anemones, which require a hard substrate to attach to, this species lives on sand flats in a tube constructed from mucus and sand.

DELAWARE

Habitats
Salt marshes and open water
Tidal shorelines and mud flats
Freshwater wetlands and ponds
Forests
Farmlands and meadows

Key Species
Snowy egret and great blue heron
Bald eagle
Black duck
Horseshoe crab
Migratory shorebirds
Blue crab
Fiddler crab
American oyster

Description
The Delaware reserve features full range tidal wetlands dominated by salt marsh cordgrass and salt hay, creeks, rivers, and bay areas, and buffeted by freshwater wooded fringe, farmlands and meadows. The reserve is endowed with a rich prehistory and a historic 18TH century plantation setting.

Location
Six miles southeast from the capitol complex of Dover, Delaware

salinity and other physical factors remain fairly constant. The sediments forming the beaches on the seaward side vary depending on their proximity to sediment-carrying freshwater sources and on the degree to which the beach is exposed to wave action. For the most part the beach is a firm mix of silt and mud particles combined with the sand. This firm bottom, lying below the low-tide mark, is home to numerous burrowing organisms, such as polychaete worms, brittlestars and clams.

One of the most abundant of these burrowing animals is the ghost shrimp, five to ten of which may live in an area ten square feet in size. Thousands of tiny holes in the beach indicate their burrow openings, with dark fecal pellets usually surrounding the entrances. Water bubbling out of the hole is a certain sign that the burrow is inhabited.

Several ghost shrimp species belonging to the genus *Callianassa* are found on Sapelo, as well as the closely aligned mud shrimp, *Upogebia* sp. The largest specimens measure six inches. Completely adapted for a life underground, they remain in their well-engineered burrow systems, which often reach six feet in length, at all times. By fanning their posterior "legs," the shrimp create water currents through their mazed burrows that pump waste material and sediment to the surface and at the same time circulate oxygen-rich water throughout the burrow.

Different species of ghost shrimp prefer different grades of sediment in which to live, and they are quite particular about it. *Callianassa major* lives in sandy beaches, *C. atlantica* inhabits mudflat and *Upogebia affinis* is found in the mud along the banks of marsh creeks. Ghost shrimp have been around for millions of years, and their burrows have fossilized very nicely in the soft sediments of ancient shores. The result is that callianassid shrimp burrows are some of the best environmental indicators of what past marine environments were like.

High-Energy Beaches

The white sand beaches on Sapelo's ocean-facing shore are its buffers against the formidable wave action of the Atlantic. The gently sloping bottom of the wide continental shelf and the shallow waters of the many offshore sandbars cause incoming waves to break several times. This releases most of the energy stored in the waves that have traveled across the Atlantic, but enough energy remains to keep the surface sediments of this shore environment in constant motion, making it a most uncomfortable and confusing place for most marine organisms.

The few animals that dwell in the turbulent and inhospitable surf zone escape their enemies and remain roughly in one place by rapidly burrowing. The most common invertebrate of this zone is the mole crab, *Emerita talpoida,* also known as the bait bug, sand bug or sand flea. Mole crabs inhabit the edge of the surf in unbelievable concentrations. As each wave passes they leap from their temporary burrows, extend their antennae, trap plankton being carried in the receding wave and then dig in again seconds before the next roller comes tumbling in. Together they move up and down the beach with each tidal fluctuation. Their round, streamlined little bodies are perfectly suited for a life among the shifting sands of a high-energy beach. Even their color—sandy gray—makes them ideally suited to their environment.

This camouflage is especially important as mole crabs are considered a delicacy by a number of other surf-dwelling creatures. Terns spot them from the sky, dive and snatch them the instant the crabs emerge from the sand. Black skimmers cruise just above the water level, their long lower bills dipping just below the surface to scoop them up. Pompano, permit, channel bass, black drum, sea trout and other surf-feeding fishes dine on mole crabs, as does the ravenous blue crab. The life of a mole crab—divided between watching the waves and watching its back—must be a neurotic and paranoid one indeed.

Another common inhabitant of the surf zone is the beautiful little coquina clam, *Donax.* As the energy of the waves constantly washes these bivalves out of the sand, they must swiftly dig in between waves to avoid being washed out to sea. Unlike their relatives living in quiet waters, clams such as *Donax* and the long, thin razor clam are very active, relying on a strong, muscular foot to rapidly dig them back into the safety of the sand. Furthermore, contrary to mud-dwelling clams, their shells are smooth and streamlined in order to make the digging process more energy efficient. Finally, surf clams differ from their mud-dwelling relatives by the presence of short, mobile siphons. Surf clams never dig in very deeply as they must move up and down the beach with the tide since sand particles cannot retain water at low tide, unlike the finer sediments of quiet embayments. The surf clam's strategy for survival revolves around being able to rebury itself quickly if exposed, while the clams of quiet mudflat habitats adopt a strategy of digging deep and staying in one place throughout their lives.

Sand Dollars and Sand Dollar Crabs

Anyone strolling along the shores of the Southeast will eventually discover the partially buried shell of another hallmark sandy-shore creature, the sand dollar. Close examination of the shell (actually called the test) will reveal a pentameric pattern of arms radiating out of the center of what is the top of the shell. This pattern is the most obvious evidence linking the sand dollar with its relatives, the sea urchins and sea stars. A living sand dollar, covered with short, moveable spines on the bottom surface, would provide even more proof that these sand-dwellers are in fact members of the phylum Echinodermata—the "spiny-skinned" sea animals.

Sand dollars do not live on the beach or even in the surf zone near the beach, although this is where we find their washed-up tests. Their round, flat bodies would be damaged by the rough action of the waves that scour the bottom of the shallow water and crash on the beach. Sand dollars are quite social, living in large communities on sand beds found in the mouths of bays, or in offshore water deep enough not to be affected by the force of the waves traveling overhead. Here the sand dollars go about their echinoid business of sorting the sand particles with their delicate spines, searching for any edible organic matter mixed in with the masses of quartz sand grains.

Within the confines of these sandy environs lives another animal, one even more limited in habitat than the sand dollar. One of the tiniest crabs in the world, *Dissodactylus mellitae* lives among the spines on the underside of the sand dollar. With claws specially adapted for hanging on to the sand dollars' spines, the crab lives, feeds and mates within this veil, the crab's universe in a sea of sand.

The sand dollar crab begins life as a tiny swimming larva, repeating a life cycle similar to that of the blue crab. After it has spent some time as a member of the plankton, the larva settles to the sea floor and hopefully lands in the vicinity of a gathering of sand dollars. How it locates its host is not exactly known. As in many symbiotic associations, the host may emit some chemical cue into the water that the crab follows, much like a scent trail in insects. However the meeting is accomplished, it is crucial to the survival of the crab that it find a sand dollar. If it doesn't it will quickly become a meal for a predator roaming the exposed habitat of the subtidal sand flat.

The sand dollar crab can be found clinging to sand dollars from Massachusetts to Florida, on three different species: *Mellita quinquiesperforata,* which lives in calm, shallow waters south of Chesapeake Bay, and *Echinarachnius parma* and *Encope michelini,* which occur in a broader range of latitude and depth along the eastern coast of the United States. When it first settles on a sand dollar, a crab may be only 0.016 inches across the carapace, or back. The young crab is small enough to hide between the spines on the underside of the sand dollar, but as it grows it no longer fits between the spines and must grasp them with its specialized claws. The tips of each claw curve and overlap, leaving a gap in the center for grasping a spine. Three of the four pair of legs are shaped like the letter "y," with the fork at the end to wedge around a spine. In this way the crab is secured to its spiny host.

These small crabs feed either directly on the loose tissue of their hosts (making them partial parasites) or in bits of organic material the sand dollar collects in its spines and grooves. Although a crab may share its home with up to seven others, rarely are more than three of them adult size. As they grow to adult size (a whopping 0.06-0.11 inches) they can and do move to other sand dollars in search of their own spiny house as their gregarious hosts "rub spines" with each other. If they do share their home it is with a member of the opposite sex. It would appear that the crabs are getting the better of the relationship—a safe home and food in the hostile environs of the subtidal sand flat. But there is little if any, negative impact on the sand dollar, making it a commensal symbiotic relationship.

Horseshoe Crabs

While the sand dollar crab may be the world's smallest, it is certainly not the most common "crab" along the eastern seaboard. This honor is reserved for one of the world's living fossils, the horseshoe crab, *Limulus polyphemus.* This striking creature has survived essentially unchanged for over 300 million years. And despite its name, horseshoe crabs are not crabs at all, but are related more closely to land-dwelling spiders and scorpions.

One of the keys to the horseshoe crabs

Sea ravens are common fish of the eastern seaboard. In an uncharacteristic flash of activity this one responded to having its picture taken by flaring its pectoral fins and leaping off the seafloor.

success is its ability to withstand substantial changes in temperature and salinity, which has allowed it to invade several coastal marine environments, from sandy beaches to estuaries to turtlegrass beds. The anatomical structure that allowed them to survive through the millennia includes a shell, or exoskeleton, that is tough but flexible. The shell provides a home for barnacles, slipper shells and tubeworms that settle on mature individuals that have grown to adult size and are no longer molting. *Limulus* can grow as large as two feet in length, including its telson or "tail," and can weigh over 10 pounds. In the adult, two compound eyes are located on the upper portion of the shell. These probably perceive movement but not actual images. On the underside of the crab are five pairs of legs, each with a claw-type pincher. The last pair incorporates a fanlike expansion and is used primarily for burrowing. The four forward pairs are used for walking and may be used for handling such eclectic prey as worms, clams, mussels and even algae. Two small pincers are used in feeding and are located near the mouth. On the inside, *Limulus* has a mouth, a gizzard-like chewing apparatus, a primitive brain and an elongated heart that pumps blue blood due to a copper-based iron-carrying pigment called hemocyanin (as opposed to our red iron-based hemoglobin). Toward the rear of the crab is the abdomen, followed by the spine-like telson. Contrary to popular belief, this appendage is not venomous, yet it still serves two functions of defense. First it provides *Limulus* with the necessary leverage required for burrowing into the sand or mud at low tide. Second, the telson can be used as a pole to flip the crab over if it is upturned.

Limulus mates in late May or June. This romantic episode always occurs at the new or full moon, when the spring tide is at its highest level. With the full tide the females can deposit their eggs high up on the beach, protected by the sand and safe from wave action. This is a marked departure from true crabs' reproductive cycles, in which they copulate and then retain their fertilized eggs on their body until the larvae are ready to be released into water. The romance begins and ends with the full moon. Each female crawls up the beach carrying a smaller male, which grasps the female's shell with its first pair of walking legs. Once up the beach the female digs 10 to 15 holes in the sand and deposits 200 to several thousand eggs in each. The male is then unceremoniously dragged across the eggs while he releases sperm. Sometimes as many as a dozen males may gather around a mating pair, each trying to fertilize the eggs, a far cry from the careful and private courting that takes place with blue crabs.

Within two to ten weeks the jellylike

eggs develop in the sand. On one of the next high tides, when the eggs are ready to hatch, the trilobite larvae struggle to the surface and head out to sea on the waves. At this point their life cycle becomes more crab-like: they go through several larval stages, molting with each one and remaining part of the plankton until they eventually sink to the bottom and take up a crawling existence as adults. Molting takes place several times a year for the first two or three years. *Limulus* becomes sexually mature at 10 to 12 years of age and may roam the eastern seaboard until the ripe old age of 20, inadvertently terrorizing summering tourists as they ramble down the beach.

Gray's Reef

Only 3 to 30 percent of the shallow and sandy continental shelf of the southeast United States is considered a hard-surface habitat. These rock outcrops are like an oasis on an otherwise sandy sea floor. One of these outcrops occurs 17 nautical miles off Georgia's sandy barrier island shores and 60 feet below the surface; it is called Gray's Reef. Because of its unique characteristics as a subtidal marine habitat it has been designated a National Marine Sanctuary, and its marine life is protected. It is host to over 500 species of invertebrates, 100 species of fish, 20 species of seaweed and is an offshore home to loggerhead turtles. The reef covers 17 square miles of sea floor and includes much of the sandstone outcrops known locally as Sapelo-live-bottom. Underwater visibility on the reef averages 25 feet, ranging from less than 10 feet to more than 60 feet. Poor visibility is mainly the result of the natural richness of the shelf waters, which contain dissolved or suspended organic material that originated in coastal estuaries and salt marshes that are part of the barrier reef system.

Blue mussels are found on both the Atlantic and Pacific coasts in cold water. They form massive beds on rocky reefs and in shallow tidepool areas, where they provide shelter for a myriad of smaller marine organisms.

Plankton that thrive on these shore-derived nutrients also contribute to the productivity of the reef. Gray's Reef waters are a mixture of sediment and organic, nutrient-laden, fresher inshore water from the estuaries and marshes of coastal Georgia, as well as the clearer, saltier inshore edge of the Gulf Stream, which is normally more than 40 nautical miles farther offshore.

Gray's Reef, unlike coral reefs composed of the stony skeletons of colonies of living and dead coral polyps, is a fossil-bearing sandstone reef formed from marine sediments. Thirteen million years ago tons of ocean sediment, including sand, silt and mollusc shells, settled on the continental shelf off the southeast coast of North America. The sediments consolidated into rock 20 to 40 thousand years ago when the area was exposed during periods of glaciation that lowered sea level. Today, the rock forms patchy outcrops along the sandy sea floor of the shallow continental shelf of the South Atlantic Bight, sticking up through the surrounding soft bottom that otherwise dominates the southeastern continental shelf.

Like the rocky reefs of Maine, Gray's Reef is covered with attached and encrusting sea life on all the exposed rock surfaces. The rock offers hard surfaces for algae and many animals, such as sponges, corals, sea whips and fans, barnacles and a host of burrowing marine organisms. These animals make up a community composed primarily of filter-feeders. In turn, this growth is a refuge for roaming reef creatures. Though the landscape of the reef appears to be dominated by encrusting animals, more mobile reef dwellers, including sea urchins, sea cucumbers, brittlestars, starfishes, snails, sea slugs, crabs, lobsters, shrimps, worms, octopods and fishes, share the available space. Even loggerhead sea turtles (*Caretta caretta*), which nest on all of Georgia's sea islands, are frequently seen resting on the bottom in cave-like overhangs of the rocky outcrops of the reef.

New marine habitats, like terrestrial ecosystems, undergo a succession of species configurations, changing rapidly at first and, eventually, becoming relatively stable and similar in composition to surrounding communities that are of the same nature. As in all habitat patches and islands, local extinctions and colonizations continue, resulting in a turnover of species. Gray's Reef may be likened to the patch of forest that is surrounded by grasslands. The life that colonizes and inhabits the forest "island" is isolated from other life forms that live in similar habitats. As a result there is more opportunity for colonization by species that are opportunistic and can fill a niche in the new habitat quickly.

Competition for resources and predation are important factors in determining the success of new arrivals at Gray's Reef. Each new arrival must secure an ecological niche in the community. This is complicated in marine animals and plants, that more often than not go through dramatic changes in their lifestyles as they mature. This includes moving from one habitat to another as do the oyster and the crab, which begin life in the plankton and end life either firmly attached to the sea floor or roaming about it.

The physical environment, especially water temperature, exerts considerable influence over the survival of marine life at Gray's Reef, and the strongly seasonal climate of the southeastern shore is an important regulator of community composition over time. In the summer, sea temperatures average 82°F at the surface and 77°F on the bottom, warm enough for tropical species to move north and colonize the reef. But during winter, sea temperatures can fall below 59°F, proving lethal to many tropical species. While mobile tropical species such as fish can migrate to warmer water in winter, it appears that the resident tropical species die off each year. These species have become annuals at Gray's Reef. The same species return again the following summer, riding on the north-flowing ocean currents that bring their floating eggs and larvae to Gray's Reef.

Fishes are the most seasonally dynamic component of the reef community. Both number of species and overall abundance are dramatically reduced during the winter months. Only the black sea bass *(Centropristis striata)*, leopard toadfish *(Opsanus pardus)*, sheepshead (*Archosargus rhomboidalis)*, belted sandfish *(Serranus subligarius)*, wrasses and groupers are common in February through March. Seasonal community changes in invertebrates are less evident. Sea anemones, sea urchins and pycnogonids (small spider-like creatures that are neither crustacean nor insect) seem to be more common on the reef in winter.

In summer, large schools of spadefish *(Chaetodipterus faber)*, scup *(Stenotomus chrysops)*, spottail pinfish *(Diplodus holbrooki)*, amberjack *(Seriola dumerili)* and scad *(Decapterus spp.)* dominate the reef. Resident reef fish include Caribbean tropicals such as the blue angelfish *(Holocanthus bermudensis)*, spotfin butterflyfish *(Chaetodon ocellatus)*, yellowtail reef fish *(Chromis enchrysurus)* and cocoa damselfish *(Pomacentrus variabilis)*. The less conspicuous and sedentary lined seahorse *(Hippocampus erectus)*, whitespotted soapfish *(Rypticus maculatus)*, crested blenny *(Hypleurochilus geminatus)* and the night-active bigeye *(Priacanthus arenatus)* are other warm-water species that inhabit the area. The comings and goings of these species are the best indicators that the South Atlantic coast is a place of constant physical and biological change.

Seafans, sponges, corals and algae are just a few of the encrusting, reef-building organisms that make up the fringing reefs of the Florida Keys. Together they create a complex habitat that supports thousands of other marine plants and animals.

Underwater Wilderness

THREE

The Florida Keys

This extreme close-up of star coral reveals a juvenile blenny trying to hide between the closed polyps. One of the most common corals in the Florida Keys, at about a quarter-inch in diameter the polyps are some of the largest on the reef.

The Florida Keys extend in a long, curving line from Fowey Rocks near Miami to the Dry Tortugas, about 65 miles west of Key West. They are low islands, rising only a few feet above sea level. Battered by towering seas and fierce winds, the Keys would seem doomed by the hurricanes that sweep up from the Caribbean every fall. However, they are protected on the west by the mainland and the shallow waters of Florida Bay and to the east and south by over 100 coral reefs and shallow lagoons, many of them named for ships that have beached during storms: Long Reef, Ajax, Triumph, Tennessee, Alligator, the Sambos, Carysfort, Molasses, French, Crocker's Hen and Chickens, Turtle Rocks, Key Largo Dry Rocks, Sombrero, American Shoal, Looe Key. The living coral reefs act as a baffle, absorbing much of the energy of storm waves before they reach the fragile keys.

Geologically, the Keys are an extension of the mainland, composed of 100,000-year-old bedrock covered with sands and muds. Nevertheless, storm surges of 8 to 12 feet are not uncommon during hurricanes, and can be expected to cover almost all of the Keys, most of which are below 5 feet in elevation.

As the glaciers formed, locking up sea water in ice, sea level dropped by more than 150 feet and the shallow platform, or terrace, was exposed to the elements of air, rain and the surf. The coral reefs died and were cemented together to form solid coral rock. The Keys became geographically isolated from the mainland beginning with the most recent rise in sea level about 12,000 years ago. At that time the sea was 20 feet below its present level. With this rise in sea level resulting from glacial meltdown, corals began to invade the newly formed sea floor, represented by the Key Largo limestone, wherever it was exposed above soft sediments and began to grow upward again to form the present reef structures.

The Keys represent the northernmost boundary for most of the plants and animals of Caribbean origins, where water temperatures are tropical year-round. This is particularly true for corals, which create the complex physical structure upon which so many other tropical Atlantic species depend.

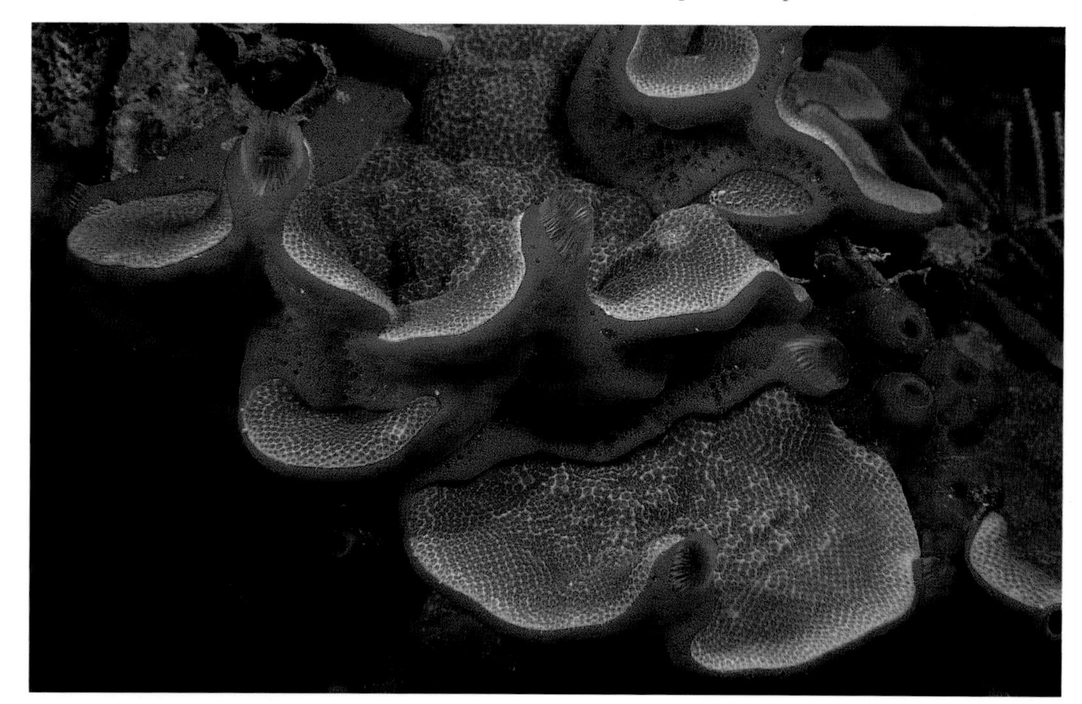

Every square inch of space is used on coral reefs. Here a sponge is growing on the underside of a small colony of corals. The sponge may actually bore into the coral, eventually eroding and even killing the colony.

Corals and Coral Reefs

Corals are animals belonging to the large phylum Cnidaria. This group of mostly marine animals includes jellyfishes, sea anemones, sea feathers, whips and fans, hydroids and soft corals. All of these seemingly dissimilar animals have one thing in common—they possess stinging cells that are used both for capturing food and for defense. The stony corals that form the backbone of the coral reefs distinguish themselves with their hard skeletons composed of thin layers of calcium carbonate secreted by the soft coral polyp, the living animal, which in form resembles a miniature sea anemone.

Each polyp is attached at the base to a tiny, limy cup that is cemented to a hard surface such as a rock or other coral. The polyp's sides and base are infolded, resulting in a cup of calcium carbonate that has inside ridges, or septa. The patterns and shapes of these ridges vary with each species and can be used as an identifying feature by coral taxonomists, who identify and describe corals. Reef-building corals are composed of a colony of polyps.

If a coral head is broken off, taken into the laboratory and sliced into thin sheets, the skeleton can then be X-rayed and the coral's growth history revealed by its annular rings. Like tree rings, they can be correlated with time, providing the growth rate and age of the coral. Studies done in Florida indicate that massive corals such as star corals (*Montastrea* spp.) grow at a rate of only about half an inch or less each year. On the other hand, corals that grow by the extension of their branch tips, such as the staghorn coral *Acropora cervicornis*, may grow as much as four inches a year.

This diamond blenny is a symbiont that lives with this common giant anemone. Like its Pacific brethren, the clown fish, the blenny is apparently immune to the stinging cells of the anemone, which provides the fish with a safe place to live.

Corals reproduce both sexually and asexually. In sexual reproduction the coral produces eggs and sperm on the swollen edges of the divisions within the polyp's body. The gender of each coral polyp within a colony may change over time. Thousands of sperm are released through the mouth of the polyp and are carried by water currents to other polyps, where the sperm enter the "female" polyp and fertilize the eggs retained in the gut.

Upon fertilization, the eggs are

Featherduster worms are conspicuous, but shy, inhabitants of coral reefs. Each crown represents a single worm, whose body is encased in a protective tube buried in the reef.

FLORIDA KEYS

Habitats
Coral reefs
Patch and bank reefs
Mangrove-fringed shorelines and islands
Sand flats
Seagrass meadows

Key Species
Brain and star coral
Sea fan
Loggerhead sponge
Turtle grass
Angelfish
Spiny lobster
Stone crab
Grouper
Tarpon

Description
The Florida Keys marine ecosystem supports one of the most diverse assemblages of underwater plants and animals in North America. Although the Keys are best known for coral reefs, there are many other significant interconnecting and interdependent habitats. These include fringing mangroves, seagrass meadows, hardbottom regions, patch reefs and bank reefs. This complex marine ecosystem is the foundation for the commercial fishing- and tourism-based economies so important to Florida.

Cultural Resources
Historic shipwrecks and lighthouses

Location
The water surrounding the archipelago formed by the Florida Keys

Protected Area
3,674 square miles

released into the water through the mouth of the parent polyp. The free-swimming larva looks like a tiny pear and is about the size of a pin head, covered with rapidly beating hair-like appendages called cilia. It may remain swimming for two or three weeks, after which the larva settles to the bottom and attaches at one end to a hard surface, such as a rock or dead coral head. The free end forms a cuplike depression surrounded by tentacle buds. The cup develops into a cavity with a mouth opening and becomes a single small coral polyp. Absorbing calcium from sea water, the baby polyp forms a hard, cup-shaped home, which represents the beginning of a new coral colony.

As the coral polyp grows upward, it reproduces asexually by budding off a new polyp, which in turn buds off another from its calcareous cup, creating genetically identical clones. This budding process forms the coral colony, and the shape it takes is genetically programmed. Ultimately, the final appearance of a coral, as with all living organisms, is the result of complex interactions between heredity and external physical and biological factors. If growth proceeds upward with lateral budding of polyps, a branching coral may develop. In other corals, the spaces between polyps are filled in with calcium carbonate and the resulting coral is solid and rocklike. Frequently, as in some brain corals, the new polyps never completely

Larger species of soft corals can grow into massive, complex colonies several feet high and wide. This candelabra gorgonian is providing shelter for a retiring squirrelfish that will emerge at dusk to roam the reef throughout the night.

separate, resulting in a long meandering "polyp" with many mouths. In some cases, environmental factors may affect coral growth so profoundly that a given coral species from one habitat may look very different from the same species living in another habitat.

The success of a coral depends on its ability to survive the rigors of its environment, and here its skeletal structure is often an important factor. Where corals may be completely exposed at low tide and subject to intense solar radiation or heavy tropical downpours, the more porous, branching corals appear to have an advantage over massive corals and can survive over three hours of exposure. Frequently such corals, especially *Acropora,* produce large amounts of mucus and appear to be covered with a clear, thick slime that apparently helps to keep them from drying out.

Shallow reef areas feel the full fury of tropical storms, which generate tremendous waves capable of dramatically damaging and rearranging the sea floor. In the wake of such storms, few, if any, corals survive either the scouring caused by rocks and sand being washed over the reef or by burial under mounds of reef rubble. However, some corals, such as the encrusting corals, are capable of surviving significant sand scouring, and massive corals are generally the last to be overturned and destroyed by wave action. In contrast, the giant branching corals, such as elkhorn and staghorn, may be totally destroyed by

Barberpole cleaner shrimp are found on both Caribbean and Pacific reefs, where they provide parasite removal services to a variety of reef fish who give them the right "high sign."

Unlike their northern cousins the Maine lobster, spiny lobsters do not sport massive claws. Rather, they use long, well-armored and agile antennae to ward off potential attackers.

JOBOS BAY

Habitats
Subtropical dry forest
Coral reefs
Fringing and basin mangrove forest
Seagrass beds
Mangrove channels
Salt flats
Lagoons

Key Species
West Indian manatee
Hawksbill sea turtle
Yellow-shouldered blackbird
Brown pelican

Description
Jobos Bay includes a chain of 17 tear-shaped mangrove islets, known as Cayos Caribe, and the Mar Negro area, which consists of mangrove forest and a complex system of lagoons and channels interspersed with salt and mud flats. The Cayos Caribe islets are fringed by coral reefs and seagrass beds, with small beach deposits and upland area.

Location
The south-central coast of Puerto Rico

a large storm. Storms serve an important funtion in distributing reefs through scatter of *Acropora* fragments.

The shapes of coral colonies may also be related to their ability to feed. For example, the branching patterns exhibited by some corals appear to influence the flow of water around and through the colony in a manner that aids in the capture of plankton. During the day, the tentacles of the coral polyp are withdrawn and the polyps are inactive. At night, the normal feeding time of most corals, the tentacles are expanded, waving slowly about in the water to capture small plankton, floating organic material and even small fishes. The tentacles possess stinging cells that contain tiny darts holding a weak poison which when fired collectively is sufficient to kill small prey, such as planktonic worms and copepods.

The growth rate of branching corals may be as much as five to ten times faster than that of massive corals, giving the branching corals a clear advantage in being able to quickly re-colonize a habitat that may have been destroyed by a hurricane. In calm water, branching corals are not subjected to damaging wave action, and here such corals grow into large, delicately branched colonies. These corals are better equipped to rid themselves of sediment, which can be a significant problem

Bristling with spines, this hermit crab makes a living sorting through sand and sediment for any living, or non-living, organic matter that might provide a meal.

in the calm waters of lagoons.

Corals have a symbiotic relationship with single-celled algae called zooxanthellae, which live in the coral tissue. This affects both the shape of coral colonies and their distribution on the reef. These single-celled plants take up carbon dioxide, a waste product of the coral polyp, and through photosynthesis produce oxygen that is used by the polyp. In addition to providing oxygen, the zooxanthellae act as a catalyst in coral skeletal growth by providing nutrients to the coral in the form of sugars, further by-products of the photosynthetic activity of the algae. Only with the assistance of the zooxanthellae can reef corals grow so massive. Of course, the algae require light to grow, and for this reason reef-building corals are most abundant at depths of less than 75 feet. Light plays a significant role in the distribution of corals on the reef and, ultimately, in their growth patterns as well. Branching corals have a much greater surface area than massive corals, increasing their ability to expose the algae to light. Many branching corals grow tall and create a significant shadow that undoubtedly inhibits the growth of other corals beneath them, thus improving their odds in the competition for space on the reef.

In addition to the physical factors affecting patterns of coral growth, there are

Flamingo tongue snails are common throughout the Caribbean Sea. This one is feeding on the polyps of a sea fan, stinging cells and all.

many biological factors operating as well. Occasionally two corals from separate colonies may come into contact with one another. When this happens, specially adapted polyps with abnormally long tentacles will lash out at the encroaching colony, stinging and damaging its tissue with these particularly virile tentacles. Ultimately the growth patterns of both corals are affected, with the more massive species frequently becoming dominant. Other organisms can also alter coral growth, including boring algae, sponges and molluscs, gall-forming crabs, shrimps that make transverse furrows on the surface of massive corals, coral-feeding molluscs, echinoderms and fishes that destroy small sections or entire colonies. These organisms give each coral head its own unique physical form and shape.

Environmental Factors Necessary for Coral Growth

Salinity, temperature and water clarity are the three primary environmental factors that control the growth and distribution of reef-building corals. The salinity must be that of full-strength sea water. Unlike oysters and blue crabs, corals are very sensitive to changes in salinity, requiring stability in the chemical make-up of the waters they inhabit.

Water must be between 68° and 86°F, with little if any fluctuation. As we have seen, the water temperature of the Atlantic changes seasonally in the northern portions, greatly affecting the range and distri-

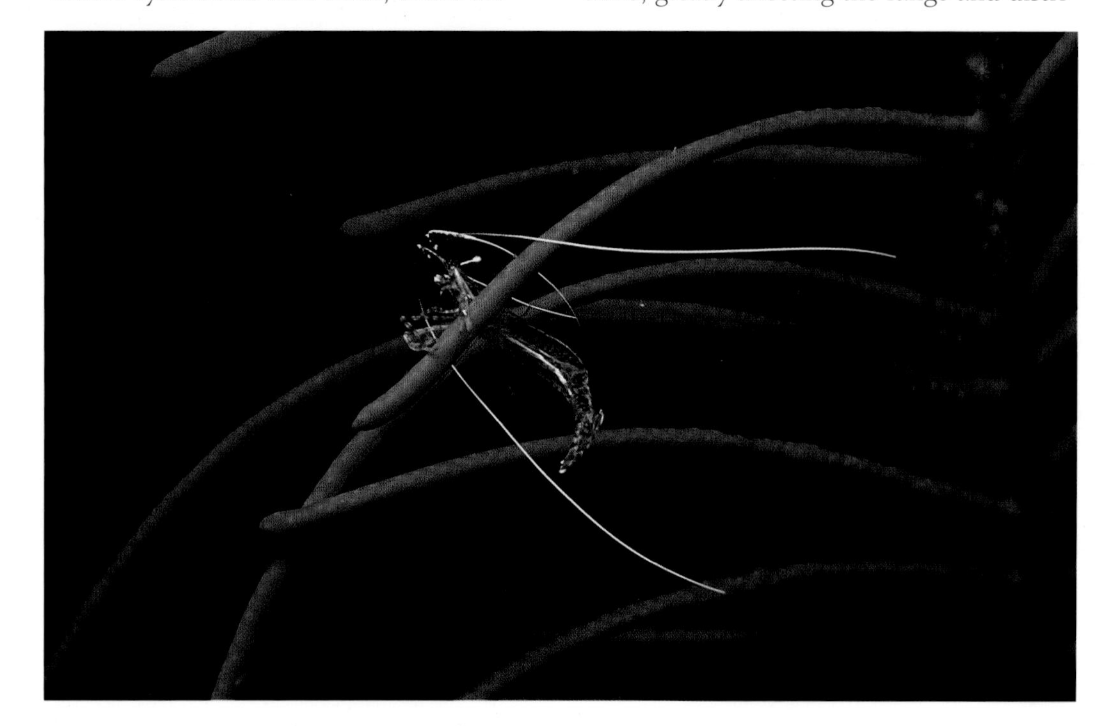

Peterson's cleaner shrimp are found singly or in small groups, usually associated with sea whips (above), anemones or corals. They sway back and forth, signaling fish that they are available to clean dead skin and small parasites from patient piscines.

bution of corals. While the Gulf Stream may move water warm enough to support coral growth under normal conditions as far north as the Chesapeake during the summer months, it does not remain long enough for even the most hardy coral species to become permanently established. The northernmost limit of coral reef growth on the East Coast is Fowey Rocks, just south of Miami. While individual coral heads are found in the Gulf of Mexico as far north as Tampa and Cedar Key in somewhat deeper water, they do not form reefs. In the Atlantic, corals are found as far north as North Carolina on the shallow shelf near the warm waters of the Gulf Stream, but only as single colonies and not as the masses that form living reef communities. This is the case at Gray's Reef. The northernmost record of coral reefs in the western North Atlantic is at Bermuda, where the corals are bathed by the warm waters of the Gulf Stream

This common Caribbean octopus is flared out in a defense display attempting to look larger than it really is. Retiring by day but active at night, octopuses are active predators on the reef, feeding on nearly anything that moves.

and the Sargasso Sea.

North of this point water clarity becomes important, as winter storms churn up the sediment on the sea floor. Heavy sediment affects corals in two ways. First, it may literally suffocate corals, which are unable to remove the sinking sediment particles from their polyps sufficiently. Second, and more importantly, corals covered with sediment are not capable of sustaining their symbiotic algae, which require light to carry out their photosynthetic activities. In Florida, where for a period of time heavy dredge-and-fill operations were employed for housing development, increasing turbidity reduced coral growth.

The Coral Reef Ecosystem

A typical coral reef is formed where corals can grow into large colonies. As the individual coral colonies grow they become crowded together. New corals grow between them, helping to fill the reef with coral skeletons. In the crevices and cracks between the corals a large variety of calcareous fragments of many kinds helps to fill in the free spaces. These are small pieces of coral, shell fragments, sponge spicules, the calcareous plates of horny corals, calcareous algal bodies, foraminiferan skeletons and many others. In some cases as much as 50 percent of the biomass of a coral reef may be made up of marine algae that act like the "mortar" holding coral "bricks" together.

Sergeant majors, damselfish family members, are one of the most common Caribbean fish. Females lay an egg mass on rocks or other exposed parts of the reef that they guard fiercely until the fry hatch.

All these encrusting organisms are slowly cemented together by chemical means. The calcareous mixture of corals and reef dwellers of other kinds are bound into the coral base rock of the reef, with only the living corals and their associates on the outer surface. Further cementation can go on to form solid rock, as in the case of the Key Largo limestone formation.

Three kinds of reefs exist in the Caribbean. The least common is the coral atoll, which occurs where a reef builds up around the shores of a small, usually volcanic island. This type of coral reef is much more prevalent in the South Pacific. Fringing and barrier reefs, which are closely related forms, are more common in the Keys. They develop parallel to the shoreline. Fringing reefs are found closer

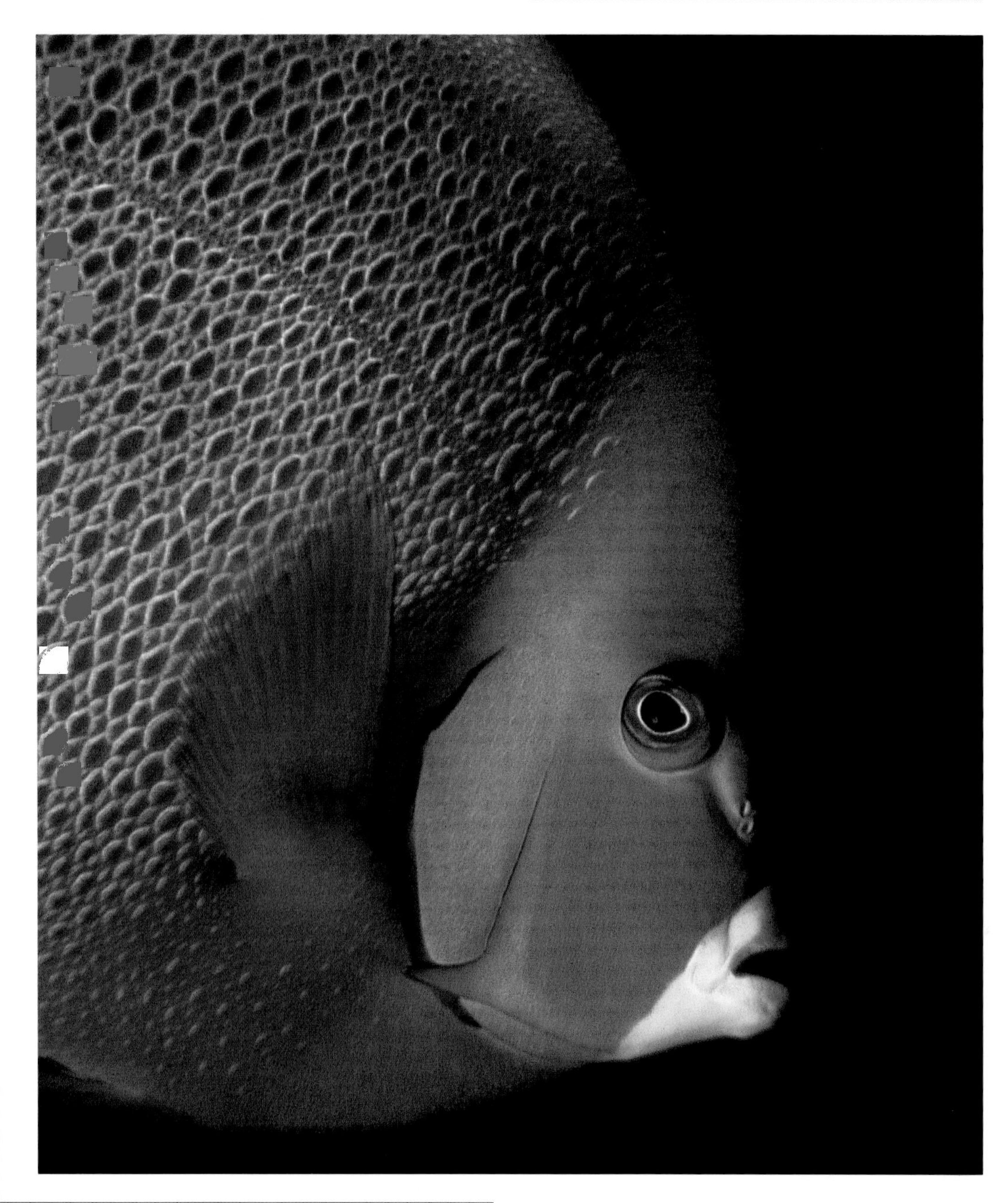

Angelfish are common inhabitants of Florida's reefs. This gray angelfish sports an opercular spine on its face in front of the pectoral fin, a defining characteristic of angelfish.

to shore than barrier reefs, which have extensive lagoons behind them. Generally these reefs grow in shallow nearshore waters in a continuum of marine life, providing shelter and food for thousands of other organisms.

Lagoon, or patch reefs, are a variation of the fringing reef and they exhibit a different structure. Sheltered by the outer reef line and surrounded by water only about 15 to 20 feet deep, these coral formations grow as isolated coral heads. The massive coral heads rise directly from the bottom, growing along the sides of the old dead coral matrix. The corals grow upward to just below the surface of the sea. The top of the reef is usually covered with staghorn coral interspersed with brain coral. Flexible gorgonians (coral relatives) cover the top in the most wave-swept zone. Unlike the outer reefs that have a solid matrix of coral skeletons honeycombed with small caverns, gorges and surge channels, patch reefs may be nearly hollow inside, their cavities having been formed by the dissolving powers of sea water.

Caribbean reefs, like those around the world, are actually thin veneers of marine life laid down upon the massive skeletons of reefs that flourished and developed in another era. In some areas, fringing or barrier reefs are very wide, such as the 5–7-mile-wide reefs off the coast of Florida. In other areas, such as off the steep, volcanic Virgin Islands, they may be no more than 100 to 300 feet wide. Regardless of its width, however, each reef can be divided into three distinct zones of life, the back reef, fore reef, and reef slope.

The Back Reef

The back reef is the first section of reef encountered from shore. It varies in width from a few yards to several miles, and in depth from a few inches to about 20 feet. Because it is so shallow, this area often experiences wide seasonal fluctuations in temperature, which can make it a rather stressful place for the plants and animals inhabiting this zone. The water in the back reef is relatively calm and silty, so that in some places thick beds of sediment accumulate. The depth and nature of this sediment largely determine the kinds of marine life present in this area.

Despite their gruesome appearance, barracudas pose no real threat to humans. They are, however, extremely effective predators of small fish which they slowly stalk, sneak up on, and strike.

Where the sediment base is thick, dense beds of turtlegrass *(Thalsassia testudinum)* develop, creating broad plains of waving grass blades. Despite this luxuriant and prolific growth, such beds often look barren during the day. However, these back reef grass beds can be very active. Huge schools of vivid blue tang roam over the grass beds, grazing much like wildebeests on the Serengeti. Huge midnight parrotfish also make excursions into the grass beds, taking huge bites out of the substrate. Again, these parrotfish might be analogous to elephants roaming the Serengeti plain. Most of the animals of the grass beds are small and cryptically colored, often green or brown. When approached, they dive into the grass to hide. All that you typically see is an occasional variegated urchin *(Lytechinus variegatus)* or a queen conch *(Strombus gigas)* wending its way slowly across the bottom while feeding. By digging deeply into the grass, however, you can also find a variety of small mollusks, crustaceans, seahorses and pipefishes. At night, life becomes more visible. Tube-dwelling sea anemones rise out of the sand and extend their long, thin tentacles into the water to collect plankton. Small snails and bristleworms *(Hermodice carunculata)* forage over the bottom. Schools of grunts and snappers, small, predatory reef fishes that spend the day hovering in the shelter of the reef, are common in the turtlegrass at night.

Where there is little sedimentation, the hard bottom of the back reef is dominated by the soft corals called gorgonians. These coral relatives also form colonies, but instead of having a completely hard skeleton their soft tissues are filled with calcium and siliceous spicules, similar in a way to the structure of a sponge. These plume-like or fanlike colonies of animals grow perpendicular to the current on flexible stalks and constantly sweep passing water with their many fine tentacles to feed on small drifting plankton. A common inhabitant (and predator) of some gorgonians is the flamingo tongue snail (*Cyphoma gibbosum*). In many parts of Florida, however, this boldly orange-spotted snail has been nearly exterminated by shell collectors who are attracted to its bright color pattern, which in life is actually part of its mantle, not its shell.

While they may look mean, moray eels are generally timid creatures, unless cornered or provoked to attack. Nocturnal hunters, they swim freely over the reef at night tracking down small fish by smell.

This area has vigorous coral growth by *Acropora* species. During storms pieces are broken off and transported to the back reef where they recement to the bottom and start growing. Only a few species of hard corals can withstand the harsh conditions of the back reef. The most common of these are finger corals (*Porites* spp.), star coral (*Siderastrea radians*), and rose corals (*Manacina* spp.). On the other hand, sea anemones and zoanthids are common on the back reef and in some areas form broad carpets of pink and green. Crabs, lobsters and shrimp are also abundant, though they are often camouflaged and can be easily overlooked. These, too, are more easily observed at night when they are out on feeding forays. A variety of fishes are also present, most of which are quite hardy and many of which are more abundant on other parts of the reef, coming into the back reef either as juveniles or occasional strays, or to feed. Aside from

Coneys are small members of the grouper family. They change color dramatically as they mature; this individual is in the "yellow" phase.

Lizardfish are highly effective sit-and-wait predators of sand channels and coral reef habitats. Lying quietly on the bottom, they simply wait for a small, unsuspecting fish to swim just a little too close to their enlarged, upturned mouths.

snappers and grunts, small angelfish, butterflyfishes, damselfishes (including the infamously territorial beaugregory, or *Eupomacentrus leucosticus*), surgeonfishes, and parrot fishes inhabit the rocky-bottomed areas. Several small sharks, including bonnetheads (*Sphyrna tiburo*) and lemon sharks (*Negaprion brevirostris*), as well as barracuda cruise the back reef feeding on unwary or injured fishes.

The Fore Reef

As you move toward open water, along the reef, you'll discover an entirely different community of reef animals. This zone is dominated by large colonies of elkhorn coral (*Acropora palmata*). This forest of living coral marks the landward edge of the fore reef, an area characterized by clear water, a gradually sloping bottom and occasionally heavy surge and wave action. The elkhorn coral zone extends vertically from just below the surface to a depth of

All puffed up and nowhere to go, this porcupinefish is displaying its typical defense response. By swallowing water it inflates its supple, flabby body, erecting its spines as the skin becomes taught.

approximately 20 feet. The horizontal width of this zone may vary from 10 to 200 feet depending upon the nature of the shore. The amber blades of the coral are massive, designed to withstand the pounding of rough seas while still providing each animal in the colony with the maximum opportunity to capture drifting plankton and sunlight. The bottom of the elkhorn zone is littered with broken blades and smashed coral colonies, the result of hurricane-force winds and waves. Swimming among these living and dead corals are colorful schools of fishes, including several species of grunts and snappers, along with many small damselfishes and butterfly fishes. Knobby mounds of star coral (*Montastrea cavernosa*) dot the bottom, mixed in with many sea fans and other soft corals. Many of these hard corals are inhabited by multi-colored serpulid worms, frequently called "Christmastree worms" (*Spirobranchus giganteus*) because of their conical shape and feathery gills.

Moving farther out over the reef and into deeper water, the stands of elkhorn coral gradually thin and give way to a barren-sandy plain. This barren zone finally ends in a jumble of multi-hued corals in which no single color, form or species dominates. Extending from depths of about 30 feet to 50 feet, this mixed zone is characterized by immense, round star and brain corals growing to over 15 feet high and 30 feet in diameter. They are often undercut at the base, producing a mushroom-shaped coral colony. The coral heads of this zone are frequently marbled with cracks and fissures, within which small reef fish such as angelfishes and damselfishes live. Some are even hollow, bored out from within by the activities of mollusks, sponges and worms. Inside the dark interiors of such coral heads you can often find animals that are more characteristic of much deeper parts of the reef. During the day these places provide hiding places for nocturnally active animals.

This mixed zone is an area of great diversity and intense biological activity due to the number and variety of habitats and microhabitats available. Competition is fierce here. Not only are the many species of corals competing with one another for space on the reef, they must also contend with sponges, hydroids, algae, tunicates and a wide variety of other encrusting organisms. Every square inch of available space is used. Dead gorgonians, for example, are commonly covered by the mustard-brown fire coral (*Millepora alcicornis*), and a wide variety of other encrusters. Crevices are packed with sea anemones, zoanthids, sea urchins and lobsters. Some of the more delicate algae also find shelter in these crevices, protected from the browsing of parrotfishes and surgeonfishes, which graze on coral and algae, respectively. Besides these common reef fishes, a wide variety of other piscine are found on this part of the reef, ranging from the pelagic jacks that cruise above its surface while feeding, to the small benthic blennies and gobies that sit on the corals and scoot around the bases of coral heads and sponges.

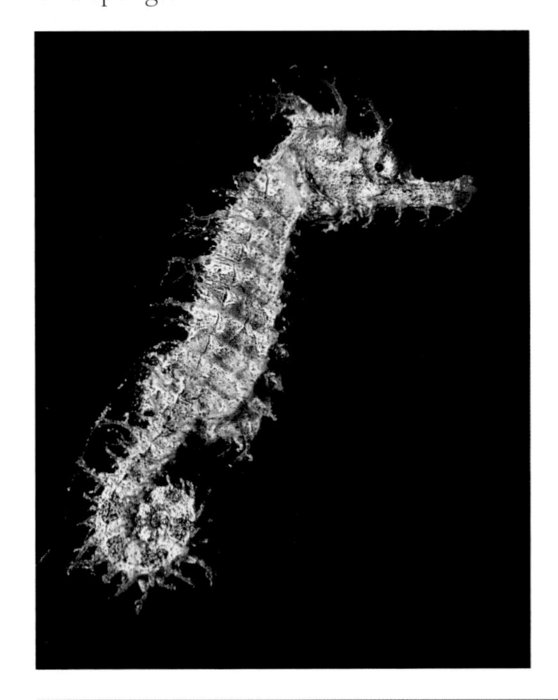

Only about two inches long, this seahorse hangs out in seagrass beds, where it blends in with the rest of the flotsam and jetsam. Camouflage is the seahorse's primary defense mechanism.

Deeper on the fore reef, the scattered coral heads characteristic of the more shallow regions of this zone begin to merge, forming immense buttresses up to 1,000 feet long that reach out perpendicular to shore. Between the buttresses, sand produced on the shallow reef pours through deep and narrow surge channels, and ultimately cascades over the front of the reef. This alternating pattern of parallel reef buttresses and sand channels is called a spur-and-groove formation. It is a hallmark of the deep fore reef zone throughout the Caribbean. The sand channels are relatively barren, containing a few isolated algal colonies and aimlessly wandering fishes. In sharp contrast, the tops and sides of the buttresses are covered with life. Coral growth in this area is particularly rich and diverse, and nearly covered by star corals, lettuce corals (*Agaricia* spp.), brain corals (*Diploria* spp.) and dense thickets of staghorn coral. These thickets extend on some reefs for hundreds of yards, the branches of the corals forming a rigid mesh in which many fishes and invertebrates find food and shelter. Along the bases of the buttresses, long tube sponges in a variety of colors extend out into the open water. These are often occupied by several small but colorful neon gobies (*Gobiosoma* spp.) that find both shelter and food in the cavities of the sponges.

The irregular surfaces of the sponges and corals that cover the buttresses provide habitats for many other reef animals as well. Brilliant orange crinoids, for example, stretch their thin feeding arms

This juvenile scrawled cowfish is perfectly formed and its colors are brighter now than it will be as it matures. Its spines and fused scales create an armored outer skeleton that protects it from attack.

into the water from cracks among the coral. The small, spider-like arrow crabs (*Stenorhynchus seticornis*) are common, as are the red-and-white banded cleaner shrimps (*Stenopus hispidus*). Lobsters and crabs are much less common in this area but are normally much larger here than they are in shallower areas. Fish life is equally diverse, with many colorful wrasses, damselfishes, gobies, and the small but brightly colored royal gramma (*Gramma loreto*) commonly seen swimming upsidedown around coral heads and feeding on passing zooplankton.

The Reef Slope

At a depth of 70 to 90 feet, the buttresses abruptly end, falling off in a nearly vertical plane known as the fore reef escarpment. At the upper edge of the drop, black corals (*Antipathes sp.*) begin to appear and reach their greatest size. Where they have been protected from collectors,

Giant of the reef, this jewfish is five feet long and weighs a couple of hundred pounds! Individuals this size are becoming harder to find as fishing pressure removes the largest "catch" from unprotected reefs.

they commonly form shaggy coral "trees" as much as 10 to 15 feet high. Unlike the true reef-building corals, antipatherians do not house zooxanthellae algae in their tissues and thus are more dominant in the deeper reef, where light levels are low and they can out-compete other corals and encrusting organisms for space. Below them, at the base of the escarpment, a barren sand flat, peppered with a few isolated coral heads, extends farther offshore out into deeper water.

Beyond the deep sand plain begins one of the most visually spectacular environments in the ocean—the deep fore reef. Starting at a depth of about 100 feet, the bottom begins to slope downward at an angle of approximately 45 degrees and then becomes more and more vertical, resulting finally in an extraordinarily eerie, blue-green cliff that is known to divers as "the wall." The wall produces instant agoraphobia, dropping to depths ranging from several hundred to one thousand feet. One gets a true sense of weightlessness swimming out into the open water away from the wall, "hanging" in what is commonly referred to as the "blue zone."

The fauna of the wall is very different from that on the shallower reef. Although many small corals dot the surface of the wall, it is dominated by great, flat plates of star and lettuce corals, as wide as 10 feet across. These cling to the cliff like shingles on a house, increasing the surface area of the colony by growing outward from a relatively small base. In this way they give their symbiotic zooxanthellae algae more opportunity to catch the dim rays of sunlight that penetrate to these depths. The undersides of the plates are coated with numerous small, delicate deepwater corals, sponges and worms that have adapted to the shadowed regions of the underhangs and caves that pockmark the wall. Around them, giant tube and vase sponges reach out into the open water. Some of these sponges are enormous. The barrel sponge (*Xestospongia muta*) for example, can grow to 10 feet tall. Such spectacular specimens are perhaps 500 years old.

Dull brown and black soft corals, which when exposed to artificial light turn brilliant red, yellow or violet, are also abundant on the wall, all feeding on the passing zooplankton.

One also finds on the wall the products of erosion and decomposition from the shallow reef far above. The sand that travels slowly down the channels between the buttresses pours through cracks along the upper edge of the deep fore reef and continually falls down the wall. Tons of white limestone platelets, formed by the brilliant green calcareous algae *Halimeda* spp., fill cracks in the coral rock and form thick beds on the upper surfaces of rock ledges. A variety of small burrowing fishes and invertebrates, many of which can be found nowhere else on the reef, make their homes in the layers of *Halimeda* plates.

While it is the invertebrates that produce the characteristic appearance of the wall, the fishes often attract the most attention. The wall not only has a unique assemblage of fishes dashing in and among its corals and sponges, but it also is regularly visited by some of the largest and most spectacular of all reef fishes. Divers visiting the wall can observe fish from two entirely different habitats: those associated with the wall and those belonging to the open ocean, the pelagic fish. Large sharks cruise the edges of the wall by day, awaiting the approach of dusk, when they will go up and over the wall's upper lip to forage on the shallow reef at night. Immense manta rays also prowl near the wall, gliding slowly through the water while filtering out the plankton on which they feed. The wall represents a transition zone where reef and open ocean meet.

The smallest member of the Caribbean grouper family, fairy basslets are abundant on deep reef walls and underhangs. This one is not impressed with the concept of being immortalized on film.

Bottlenose dolphins abound on the Florida Gulf coast, where family groups hunt, mate and play in the shallow waters that extend for miles offshore. Known best as "Flipper," this species is the most common small, gulf cetacean.

FOUR

The Gulf of Mexico

Mixed schools of bar jacks (middle) and "grunts" are not uncommon in the Gulf or throughout the Caribbean region. From a distance their shape and color combine to create a homogeneous mass of silver confusion to predators such as barracuda.

Along the shores of the Gulf Coast, you'll find everything from coral shores, alluvial plains, limestone plateaus, sand dunes, tide flats, hypersaline bays, beaches, lagoons, mangrove ridges, barrier islands, deltas and bayous, mudflats and salt marshes.

Some types of coastline here are considerably more ancient than others. While some are dominated by the continent, others are dominated by the sea. All inshore waters are shallow as a result of a wide shelf that extends out to the 100 fathom mark, in some places more than 100 miles offshore. Even at its center, the Gulf lacks the great depths of the Atlantic Ocean. Because generalizations are impossible for this conglomeration of habitats, several of the more prominent and interesting types of shoreline and continental shelf associations of life are described below. It must be remembered, however, that the difference—measured perhaps by only a few miles—between a sandy beach ridge and a delta's muddy alluvial fan are so great that comparisons are difficult to make.

The Geology of the Gulf

The western Gulf, from the Mississippi delta to the Campeche Canyon on the west side of the Yucatan shelf, illustrates clearly the influence of delta sedimentation on the continental shelf of the Gulf. The Mississippi River, because of its great drainage area, has been the major contributor of sediments and has built an extremely large coastal-plain/continental-shelf complex in the northwestern Gulf of Mexico. The narrowing of this shelf southward into Mexico is due to the limited drainage areas of the rivers and streams that flow into that part of the Gulf—there simply is not as much sediment flowing onto the continental shelf in this area.

On the other hand, the West Florida and Yucatan shelves are broad, shallow areas made up mostly of carbonate sediments produced by lime-secreting organisms such as corals, calcareous algae, tube worms, bryozoans and other encrusting invertebrates. These areas of little or no continental sediment are underlain by thousands of feet of shallow-water sediments of organic origin. This indicates that the accumulation on the sea floor of carbonate skeletons from tiny, mostly single-celled organisms in these areas has kept pace with the sinking or subsidence of the region for many tens of millions of years.

Another structural complication in the northwestern Gulf is the formation of huge salt domes on the outer continental shelf and continental slope. These are remnants of what was an enormous plain of salt, created at a time when this part of the Gulf was exposed to air. The salt accumulated as a result of evaporation over millions of years. When the area was submerged again the salt was buried by miles of sediments dumped into the Gulf. Some of it migrates upward, lifting the seabed. Two of these salt domes, known as the West and East Flower Garden Banks, are unique in that they are capped with the northernmost coral reef structures in North America.

Geography and Environmental Features of the Gulf

The Gulf of Mexico, a body of water covering approximately 615,000 square miles, is very nearly an inland sea. It is 1,000 miles across at its widest point. The entrance from the Caribbean is the 100-mile-wide Yucatan Passage, pinched together by Florida and the Yucatan Peninsula of Mexico. Cuba lies directly between the two. Water leaves the Gulf

ROOKERY BAY

Habitats
Mangrove forests
Barrier beaches
Coastal dry-zone scrub
Pine flatwoods
Seagrass beds
Tropical hardwood hammocks
Open shallow waters

Key Species
Red, white and black mangroves
Bottlenose dolphin
Manatee
White ibis

Description
Rookery Bay features pristine mangrove forests surrounding shallow bay waters. The upland buffer consists of pine flatwoods and dry-zone scrub. Resource management efforts include prescribed burns, exotic plant removal, restoration of disturbed sites and marine mammal recovery and rescue.

Location
5 miles south of Naples, Florida

through the Straits of Florida, which open into the Atlantic Ocean. Florida, Alabama, Mississippi, Louisiana and Texas line the eastern, northern and part of the western shores. Mexico completes the remainder. Although much of the Florida mainland is a limestone plain, most of its shoreline from the southern tip (including the Keys) and north along the west coast is of living origin. This includes patches of ancient and modern coral growth and broad expanses of mangrove swamps, including the multitude of mangrove islands in the Ten Thousand Islands region of western Florida. The remainder of the U.S. Gulf Coast is composed of alluvial sandy and muddy sediments derived long ago from ancient rivers and glaciers. Barrier islands border much of the Gulf, creating additional habitats. As on the East Coast, they protect the continental shorelines behind them from the potentially devastating effects of storms.

The Mississippi delta juts out from the northern shoreline of the Gulf. The Mississippi's dynamic, ever-growing deposit of river sediments provides great biological potential and opportunity. In addition to the billions of gallons of fresh water entering the Gulf daily, the Mississippi brings two million tons of silt every day to the Gulf—representing material from nearly half the continent, including chemicals and industrial wastes from hundreds of cities and towns, and fertilizers from thousands of farms. There is no question that this giant of a river has a tremendous effect on the entire Gulf region. The ceaseless introduction of minerals from the continental U.S. is a major source of nutrients for the Gulf, resulting in rich plankton growth. In addition, there is an enormous transfusion of organic material from the decomposition of marsh grass and other coastal plants from the Mississippi delta and the vast marshes lining the Gulf's shores.

Deltas are the products of rivers. As river velocity decreases and sediment drops to the bottom, this material builds the delta both outward and to the sides. River sediments drop out of suspension quickly in the Gulf of Mexico, for less than a hundred miles out from the Mississippi the water is as clear as anywhere in the world. The discovery of an ancient, now-buried channel thirty miles from the present river mouth suggests that the river has not always emptied into the same spot in the Gulf.

Mangrove swamps, coral shores, sandy beaches and salt marshes are all coastal habitats present in the Gulf. Each constitutes only a fraction of the entire Gulf Coast, and two—coral reefs and mangroves—are restricted almost wholly to the southwestern coast of Florida and the Yucatan Peninsula.

The extensive system of Gulf barrier islands, like those along the mid-Atlantic coast, serve as obstacles to storm waves, although frequent hurricanes may wash over them with great destruction to vegetation, wildlife and developments. Barrier islands result from the windward effects of wave and current forces in the sea.

Climates, Tides and Water Circulation

The Gulf is known for its hurricanes, four-fifths of which form far outside of the region. They enter the Gulf usually from the southeast. These enormous storms, which may range up to 500 miles in diameter, cause most of their damage by generating wind-driven tides and waves that crash into vulnerable inshore areas across breakwaters and seawalls. The low barrier islands of the Gulf of Mexico are particularly susceptible to these storms.

Except for storm tides, tidal fluctuations in the Gulf of Mexico are not impressive. In most places, the rise and fall amounts to no more than about one or two feet. Frequently the effect of the tides is magnified by the extremely gradual slope of the shore. The actual shoreline remains much the same from high to low tide, without the wide intertidal zones seen along the Atlantic and Pacific coasts.

Winds, barometric pressure changes and the flow of many large rivers each affect the rise and fall of the tides in the Gulf. Onshore winds push water against the coast, and differing atmospheric pressures turn the Gulf into a huge barometer as water levels rise and fall. River runoff changes with the seasons. Melting snow and ice produce greater volumes, but during late-summer droughts, smaller amounts of runoff are the rule. River flow alone can have a marked effect upon water level along the coastal regions of this enormous, nearly enclosed sea. The comfortable, predictable intertidal rhythm

APALACHICOLA BAY

Habitats
Forested flood plains
Open water
Oyster bars
Salt marshes
Barrier islands
Freshwater marshes

Key Species
Tupelo gum
Bald cypress
American oyster
White shrimp
Osprey
Black bear
Atlantic sturgeon

Description
Apalachicola features 1,162 vascular plant species, 315 species of birds, over 180 species of fresh, estuarine, and salt-water fish and 57 species of mammals. The highest species diversity of amphibians and reptiles in North America north of Mexico has also been listed from the Apalachicola Bay basin.

Location
55 miles southeast of Panama City, Florida

of the Atlantic shore is missing from the Gulf of Mexico.

The Gulf is an almost circular basin with only two narrow openings to the ocean. The currents in the Gulf are complex. In general terms, water enters through the Yucatan Channel, circulates into the region of the central Gulf in a pattern called the "Loop Current," then leaves through the Florida Strait. Spin-off eddies and coastal currents add considerable complexity and variability to the Gulf's water movement. Warm water flows northward into the Gulf from the Caribbean, which, in the summer, is further heated by the sun acting upon the broad expanse of shallow nearshore areas, which may be only 20 feet deep. At this time of the year, surface temperatures range between 80° to 90°F. Conversely, nearshore water temperatures can drop significantly in the winter when northern air masses descend upon Gulf shores. Some bay waters have actually been known to freeze in extreme cases. These fluctuations in temperature, combined with seasonal variations in fresh water entering the Gulf, make its coastal marine environments hospitable only to highly adaptive species of organisms. Offshore, temperature and salinity vary much less due to the active mixing of water masses.

Surface currents continually enter the Gulf of Mexico from the Caribbean Sea through the Yucatan Channel, a relatively narrow passage about 130 miles wide. In the summer this tropical water in the Loop Current is widely diffused into the northern Gulf, but in the winter it tends to circulate somewhat farther to the south. The cause of this seasonal difference is undoubtedly related to the frequent occurrence of northerly winds during the winter. These winds also lower the surface temperature of the nearshore water until it is below the tolerance of most tropical

WEEKS BAY

Habitats
Tidal flats
Seagrass meadows
Salt marsh
Forested swamps
Upland forests

Key Species
Black neddlerush
Brown pelican
Great blue heron
Shrimp
Sea trout
Blue crab
Redfish
American alligator

Description
Weeks Bay is a small embayment that is a critical nursery for fish and shellfish. Weeks Bay is a shallow subestuary of Mobile Bay and is located in Baldwin County, receiving water from the Fish and Magnolia rivers. The watershed is ideally sized for estuarine research purposes.

Location
U.S. Highway 98 between Mobile, Alabama and Pensacola, Florida

organisms that may have migrated into the area from the south.

With such marked seasonal current and water temperature changes, it is not surprising that considerable changes occur in marine life. Along the Texas coast in the latter part of April, when the average water temperature climbs to about 68°F, there is a significant change in the fishes you'll find here. Some cold-loving species migrate offshore into deeper water while others that are apparently attracted by the higher temperature will move inshore. Still others will migrate into the area from the south, most probably from the Gulf of Campache.

The Florida Current is considered a part of the extensive Gulf Stream system, and carries the outflow from the surface of the Gulf of Mexico. It is a swift stream that extends eastward through the Straits of Florida and then turns north towards Cape Hatteras. as part of the Gulf Stream. It is reinforced by the Antilles Current, a stream that comes from the east side of the Greater Antilles, having originated from the North Equatorial Current. The Gulf Stream has a profound effect upon the biota of the Atlantic Coast, from northern Florida to North Carolina. It carries so many tropical organisms into the area that their numbers have tended to obscure those that are permanent residents. There are even a few tropical species that are able to withstand the winter temperatures of onshore waters.

Marine Habitats of the Gulf

The Gulf's size, physiography, geology and geographical extent result in a number of different marine habitats. The eastern and western shores of the 4,000-mile-long coast share similar types of marine communities and sea life but the habitats, plants and animals found along the north and south shores of the Gulf are quite

different. The Gulf of Mexico is so huge it encompasses both temperate and tropical regions. The transition is more pronounced in coastal habitats and species than offshore.

Coastal Habitats

Along the coast, temperate salt marshes dominate in the northern Gulf from Tarpon Springs, Florida, to at least Port Isabel, Texas. Populations of tropical black mangroves intermix with salt marshes as far north as the Mississippi Sound. Well-developed tropical mangrove forests dominated by red and black mangroves replace sal tmarshes southward from Tampa Bay and Laguna de Tamiahua in the Gulf. They are not significant in the northern Gulf .

Temperate oyster reefs are extensive in the northern Gulf from Cedar Keys, Florida, to Port Aransas, Texas. Reduced reefs occur in Gulf estuaries as far south as Laguna de Terminos, Campeche. Tropical seagrass beds are abundant inshore throughout the Gulf except between Galveston and the Mississippi delta, where they are limited by high turbidity and low salinities. Beds in the Mississippi Sound may be restrained reproductively by temperate conditions. Jetties in the northern Gulf bear biota similar in many respects to those of jetties in the Carolinas. Tropical forms are more abundant on south Texas jetties such as at Port Aransas and Port

This rock hind feeds on small fish, shrimp and other small animals living on the Flower Garden Banks. As with so many of the "Banks" inhabitants, the rock hind's range extends throughout the Caribbean.

Isabel. Additionally, south Florida intertidal rocks support assemblages that are distinctly more tropical than temperate in nature.

Subtidal Habitats: Soft Bottoms

Subtidal communities of the Gulf are represented by three general habitats, characterized by substrate type: sands, muds and hard banks. At the same time there are innumerable combinations of sediment. For example, near river mouths along the northeastern Texas coast, sand/mud mixtures are common and considerable debris (including rocks, logs and garbage) is washed seaward and deposited offshore. Further, a number of microhabitats can be differentiated according to relative proportions of sands, muds and debris found on any one part of the Gulf sea floor. Such distinctions become less pronounced toward southwestern Texas shores, where the sea floor is more uniform.

A secondary, but important, distinction can be made between the Gulf's subtidal sediments based on their origin. Land-based sediments (primarily clay and quartz) house distinctly different kinds of organisms than do organically derived sediments made of coral and shell fragments. The floor of the Gulf can be mapped according to these two main sediment groups. The differences between these substrates is so clear that they can also be named according to the kinds of organisms found on each.

The "white shrimp" grounds (inner shelf) and "brown shrimp" grounds (outer shelf) are present throughout the Gulf where land-based sediments predominate. White shrimp grounds are best near hyposaline estuarine environments such as the stretch of ocean between Pensacola and Texas and off Laguna de Terminos. Offshore, brown shrimp grounds extend from Pensacola to the Gulf of Campeche. Both are most extensive off Texas and Louisiana. Throughout their range the shrimp communities are fairly uniform, and geographic ranges of the species extend substantially northward toward the Carolinas and southward in the Caribbean.

Pink shrimp grounds are restricted to organically derived sand bottoms on the Yucatan and West Florida shelves, and are best just north of the Tortugas and off southern Campeche. Tropical communities occupy the continental shelves of Yucatan and West Florida, extending as far north as the mouth of the DeSoto Canyon off Pensacola. They are held offshore north of Cape Romano, Florida, by coastal land-based sands and increasingly temperate coastal conditions. As mentioned earlier, the diversity of tropical species on these vast areas of sediment decreases the farther north you go.

Perhaps the most colorful fish in the Caribbean, queen angelfish are a familiar sight on reefs throughout the region, including the Flower Garden Banks and the hard-bottom reefs of the Florida Gulf Coast.

Colorful, active and efficient predators of the Flower Garden Banks, marbled groupers play an important role in regulating population densities of smaller fish species on these subtidal "islands" of diversity in the Gulf.

Hard Banks

Throughout the Gulf are scattered hard carbonate hills, or banks, that rise above the surrounding sediment plain. The banks off the Texas and Louisiana coasts apparently originated at a time when sea level was considerably lower than today. These "relic habitats" provide homes for organisms not otherwise found in the region. As you move southward along the continental shelf, the banks become more prominent and eventually rise to the surface as emergent reefs of tropical Mexico on the western shore, and Florida reefs on the eastern shore of the Gulf. Specifically, tropical coral reefs occur southward from Miami, the Dry Tortugas and Cabo Rojo, Mexico. Off eastern Florida, tropical reef organisms diminish north of Miami and are replaced by more temperate submerged banks off St. Lucie Inlet. A few hearty reef-building corals occur as far north as the Carolinas. In the Gulf, tropical "hard grounds" extend northward to the Florida Middle Ground on the West Florida shelf. A few submerged coral reefs and deep-water tropical reefs are present on shelf-edge banks off North Texas and Louisiana. Other deep-water tropical reefs are found on the outer southwest Florida shelf south of Charlotte Harbor and in the Atlantic off Key Largo. In the northwestern Gulf, nearshore hard bottoms bear basically temperate plants and animals that may have a mix of some tropical fishes and invertebrates, such as shrimp and crabs. The oil platforms off Texas and Louisiana, over 3,500 in all, also provide artificial hard substrates that are encrusted with tropical organisms near the shelf edge, mixed tropical and coastal sea life on the mid-shelf and basically temperate organisms with a few tropical forms nearshore. During the summer some tropical reef fishes move inshore throughout the northern Gulf, occupying coastal flats and hard grounds.

The Flower Garden Banks

While there are coral reefs in the most southern portions of the west and southeast margins of the Gulf, there are no *nearshore* coral reefs north of 22°N latitude. Coral reefs do occur, however, on two unique promontories lying near the edge of the continental shelf somewhat more than 100 nautical miles south southeast of Galveston, Texas. Named the Flower Garden Banks, each bank caps a salt dome rising above the surrounding land-based sediments of the continental shelf. At approximately 28°N latitude, the Flower Gardens contain the northernmost coral reef communities on the North American continental shelf. Only the Bermuda Islands, bathed in the warm Gulf Stream waters at about 32° 20′ N latitude in the Atlantic, have coral reefs at a more northerly location than those of the Flower Gardens. These reefs are truly unique, and have been designated as one of the country's national marine sanctuaries. To put their position in perspective, the nearest reefs to the Flower Gardens is on the Yucatan shelf about 442 miles to the south. The reefs nearest in latitude to the Flower Gardens occur off southern Florida, over 693 miles away.

Both the East and West Flower Garden Banks crest at a depth of about 60 feet. The West Flower Garden Bank is a long ridge extending northeast to southwest. The shallowest part of the bank extends east to west for about three-fourths of a mile at about 90 feet. This shallow section varies considerably in width, being widest (about 1,140 feet) somewhat east of center. Smaller in size and slightly shallower on average, but more recently studied, the East Flower Garden Bank is a pear-shaped dome lying about 7.5 miles east of the West Flower Garden Bank, covering about 42 square miles. Slopes are steep on the east and south sides of the bank and gentle on the west and north sides. The water surrounding both banks varies in depth from 300 to 450 feet.

Over the past thirty years, the Flower Garden Banks have been studied extensively. These coral reefs are actually growing on bedrock that is lying on the tops of salt domes which represent remnant portions of rock that was broken up as these enormous salt deposits began to erode. Seismic data indicate that the top of the salt may be within only 90 feet of the crest of the reef.

Ecology of the Flower Garden Banks

Positioned out on the edge of the continental shelf, the Flower Garden Banks are protected from the sediments pouring into the Gulf from rivers like the Mississippi, and are not affected by the temperature extremes characteristic of the shallow coastal waters nearshore. Other hard banks of the northern Gulf are chilled by winter temperatures which regularly fall below the critical minimum temperature required for most reef-building corals (61 to 64°F). Corals on the isolated Flower Garden Banks are also protected from scouring by sands swept along the bottom by strong nearshore currents, nor are they subject to the sediment-laden, turbid water conditions found near the coast. These are the primary reasons why reefs are mostly absent on the hard banks of the northern Gulf region. The hardy coral species, such as *Oculina* sp., is occasionally found nearshore, but, for the most part, large communities of reef-building corals are absent.

Because of their distance from shore, the Flower Garden Banks are not strongly affected by seasonal changes in weather

FLOWER GARDEN BANKS

Habitats
Coral reefs
Algal-sponge communities
Brine seep
Sand flats
Pelagic, open ocean
Artificial reef

Key Species
Star coral
Brain coral
Manta ray
Hammerhead shark
Loggerhead turtle

Description
One hundred miles off the coasts of Texas and Louisiana, a pair of underwater gardens emerge from the depths of the Gulf of Mexico like oases in the desert. The Flower Garden Banks are surface expressions of salt domes beneath the sea floor. This premiere diving destination harbors the northernmost coral reefs in the United States and serves as a regional reservoir of shallow water Caribbean reef fishes and invertebrates.

Location
Roughly 110 miles south of the Texas-Louisiana border

Protected Area
56 square miles

and oceanographic conditions that play such a large role in shaping the nearshore marine life along the northern Gulf coast. Instead, these oases of marine life are continually bathed by warm water from the southern Gulf basin nearly year-round. Surface water temperatures rarely drop below about 68°F and often reach 86°F in the summer. The banks also emerge out of the sea floor in comparatively deep water, and these two aspects combine to insulate them from drastic changes in water temperature. Additionally, water surrounding the banks is clean and clear oceanic water. Visibility regularly exceeds 100 feet around the Flower Gardens. The oceanic water flowing over the banks is sufficiently clear and the crests of the banks are sufficiently shallow to encourage the growth of reef-building corals which are dependent on the photosynthetic activities of their symbiotic algae partners. Finally, the southern tropical oceanic water provides a continuous supply of plankton produced by tropical "parents" originating in the Caribbean and tropical Atlantic. Which species survive the journey and can establish themselves is partially a function of individual capabilities to cope with a reasonably long journey at sea and the abilities of the settled adults to cope with the physical conditions near the extreme edge of their distribution ranges.

While the coral reefs growing on the East and West Flower Garden Banks have essentially the same marine life, they differ significantly from the coral reefs of the Caribbean and tropical Atlantic. One of the most apparent differences is the lack of any significant soft coral populations at the Flower Gardens, organisms that are major players in Caribbean reef ecosystems. Apparently the planktonic larvae of sea fans, sea whips and other similar octocorals simply cannot make the trip from distant tropical reefs. While the environmen-

tal conditions for reef growth may be adequate at the Flower Gardens, there is still the need for a source of larval recruitment to populate them. In general, species diversity at the Flower Gardens is considerably less than that on reefs of the central Caribbean. For example, about 20 species of hard corals occur on the Flower Garden Banks, but the reefs of Jamaica, Roatan or Grand Cayman are home to about 50 species. There are approximately 80 species of algae, 250 species of invertebrates and 175 species of fishes living on the banks—these are the ones that have been successful at colonizing these remote outcrops of rock.

Zoogeography of the Gulf of Mexico

Between the coral reefs of the Flower Gardens and the Yucatan shelf and the shrimp grounds off Texas, a wide variety of marine communities are found in the Gulf. This diversity of habitats and marine life has generated numerous schemes by marine zoogeographers over the years. The most accepted scheme puts the northern Gulf within the warm-temperate Carolinian province, which also includes areas of the southeast Atlantic coast. The marine life of the northern Gulf has become isolated by Florida from similar shores along the eastern seaboard from Virginia to Georgia. This isolation has existed for many thousands of years, producing what is considered a "relic" northern Gulf community. The degree of endemism in this area is not extensive, evidence that it may be in its early stages of evolution as a distinct zoogeographic region or that larval exchange with the East Coast is still occuring.

Species that are more typical of the cold-temperate fauna of the East Coast can also be found in the Gulf. Some species occur year round while others are seasonally present or their numbers vary with the time of year and corresponding changes in weather and oceanographic conditions. These seasonal changes in the character and make-up of Gulf marine flora and fauna are most evident in the northern Gulf, creating temporally diverse communities in contrast with the increasingly diverse shores of tropical Mexico and the Florida Keys.

Three primary physical factors have determined the distribution and constitution of the Gulf's marine life: water temperature, substrate type and the flow of water currents in the Gulf. There appears to be a fairly well-defined north-south boundary between the tropical and warm-temperate groups of marine life found in the Gulf of Mexico. This border extends across the Gulf from Cabo Rojo, Mexico, to Cape Romano on the west coast of Florida. The area north of this line is dominated by warm-temperate species and is visited periodically during the summer by tropical forms that swim (or float) in when water temperature rises. The winter water temperature here may occasionally get below 50°F in the shallow bay shelves of north coastal waters. The effects of seasonal drops in temperature diminish as one moves south on both sides of the Gulf and tropical to subtropical conditions are the norm on the outermost Gulf shelf.

Brain coral can form massive colonies in areas of high surf and wave energy, where they can outcompete more fragile forms of coral for space.

The low winter coastal temperatures exclude most—but not all—of the tropical species originating from the south. This creates a basically warm-temperate coast characterized by habitats such as marshes and oyster reefs but also having significant tropical habitat representatives in black mangroves and tropical seagrass beds. The appearance of temperate species of algae in the winter and tropical fishes during the summer on coastal and nearshore hard bottoms and artificial structures such as jetties and artificial reefs is also characteristic of this transitional and variable part of the Gulf. These environmental conditions extend south to at least Tarpon Springs in Florida and the Laguna Madre de Tamaulipas in Mexico. By the time one has gotten to Ten Thousand Islands, Florida, and Cabo Rojo, Mexico, the transition to primarily tropical forms is fairly complete.

A direct correlation exists between the distribution of offshore benthic marine life and the nature of the substrate on the sea floor. Substrate type is the primary physical factor influencing the distribution of marine benthic communities offshore

In terms of their sediment make-up, the offshore bottom communities of the Gulf show a distinct change from east to west, with the influence of the Mississippi River exerting itself on the continental shelf. On the eastern side of the Gulf the bottom is characterized by substrates of carbonate origin that generally support elements of tropical benthic marine life although this element decreases gradually toward the north as water temperatures become cooler and more variable seasonally. There is a virtually continuous association of coral patches and sponges covering the broad continental shelf west of Florida from the Keys almost to the western boundary of the state. West of this point, the coral-sponge association is abruptly replaced by vast undersea plains of terrigenous sediments of sand, silt and clay dumped into the Gulf by the rivers emptying into it. These soft bottoms support the brown and white shrimp communities, which reach their greatest degrees of abundance and diversity on the mid-and outer shelves of the northwestern and southern Gulf of Mexico.

The movement of water into, through and out of the Gulf has a tremendous affect on the kinds of marine organisms found there and how they are distributed. This is most evident in the tropical species present throughout the Gulf, which ride in on the tropical waters entering the Gulf from the Caribbean Sea. As this warm tropical water moves through the Gulf it mixes with fresh water entering the Gulf along the north shore, modifying both its salinity as well as temperature as it comes into contact with cold winter air temperatures. Depending upon the time of year, this huge water mass may be affected by these events as far south as Miami. Otherwise, the immense open water currents of the Gulf are essentially tropical, and the agent of dispersal of tropical larvae to any appropriate substrate in areas having the right salinity, temperature, light and turbidity conditions to allow for their growth and development. The successful coral reef assemblage on the Flower Garden Banks is a clear example of the influence these current patterns have on the distribution of marine life in the Gulf.

This pair of chromodorid nudibranchs (sea slugs) are getting to know each other in a quiet backwater of a Florida Gulf Coast mangrove forest.

Relationship of the Atlantic and Gulf Coast Faunas

One of the most striking aspects of the zoogeography of the Gulf of Mexico is the relationship of its marine flora and fauna to that of the Atlantic coast. For example, about 85 percent of the Atlantic shore fishes can also be found in the Gulf. The other 15 percent are species that invade the Atlantic coast from the north. Additionally, there are few endemic species found along the warm-temperate part of the Atlantic coast, indicating that the vast majority of species found there have their evolutionary roots somewhere else. It would appear that many mid-Atlantic marine species originated in the Gulf of Mexico long ago. The question is, how did they get there?

We have already determined that water temperature has a direct bearing on the latitudinal movement and distribution of marine life in the Gulf as well as along the eastern seaboard. In theory, the warm-temperate species in the northern Gulf are isolated from their East Coast brethren, the southern portion of the state of Florida acting as a physical barrier that effectively separates the two components of what seem to be related faunal groups. As a result of this isolation one would expect to find some evolutionary divergence between the two assemblages, Gulf and Atlantic. However, the low rate of endemism on the Atlantic coast does not suggest this. From a zoogeographic standpoint, the very high degree of relationship to the Gulf Coast and the almost negligible amount of endemism are good indicators that the two warm-temperate faunas must have been in communication a relatively short time ago and may still be.

If Florida acts as a physical barrier to the exchange of genetic material between the two assemblages, then perhaps an exchange occurred during a time when this low-profile peninsula was submerged during a period of global warming and a concomitant rise in sea level. The last time this happened was during the Pliocene about a million years ago. However, there is so much similarity between the two groups that there must have been—or still is—an exchange of genetic material going on. These similarities in species composition could not have been maintained this long without substantial transfer of genetic material in one direction or another. So while the submergence of the Florida peninsula may have allowed for some exchange, changes in global weather patterns, the rise and fall of sea level, and the lowering of ocean water temperatures may also have played a role in the movement of species around the Florida peninsula. During periods of glaciation, sea level drops as oceanic water becomes locked up at the poles and global ocean temperatures decrease as well. This would allow warm-temperate species from the Gulf to travel around the peninsula during these periods. The patterns of the currents in the Gulf, the Straits of Florida, and the Gulf Stream, which move great volumes rapidly out of the Gulf and up the Atlantic coast, strongly favor the transfer of species by larval transport from the Gulf to the South Atlantic and not the other way around. Evidence supporting this idea includes the significantly greater diversity of marine life in the Gulf and its substantially greater degree of endemism (approximately 10 percent). These faunal features indicate that the center of distribution and evolution of these assemblages is the northwestern Gulf of Mexico, not the mid-Atlantic coast, and that they should be considered within the same faunal province, namely the Carolinian.

There is still yet another explanation for the transfer of species around the Florida peninsula which may be happening today. This is the idea that the larvae of some warm-temperate species such as penaeid shrimp can tolerate tropical waters for brief periods of time—enough to make it around the peninsula and up the East Coast. Species such as the brown and white shrimp may not be found off the west coast of Florida not because of higher water temperatures but because they simply cannot deal with the carbonate substrates that dominate this coast or compete with other organisms that are already established there.

To sum up, it can be said that the Carolina region is a distinctive warm-temperate area consisting of two closely related but geographically separated areas. It includes the northern Gulf of Mexico, north of Cape Romano in the east and Cabo Rojo in the west. The Atlantic coast area extends from Cape Kennedy to Cape Hatteras. Historically, the northern Gulf has probably been the primary evolutionary center for the Carolina region and it has contributed species to the Atlantic coast section via a migratory route around the Florida peninsula which was facilitated by a general lowering of global ocean water temperatures during glacial periods. The northern Gulf is richer in species and also demonstrates a relatively high degree of endemism (about 10 percent) in both invertebrates and fishes when compared with the Atlantic area. If one considers both areas of the region—the Gulf and the Atlantic—endemism is quite high, probably about 30 to 40 percent.

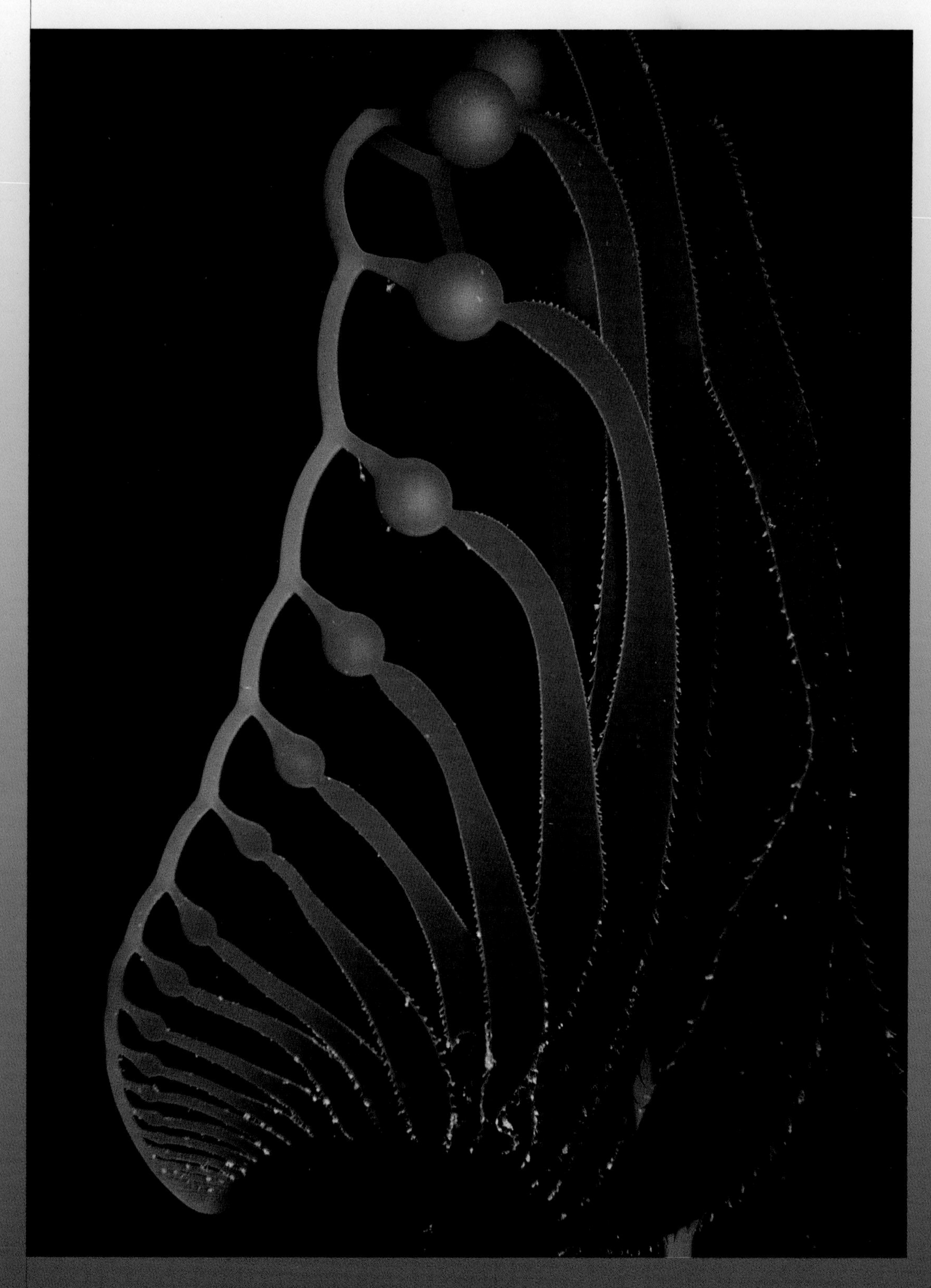

Growing at up to a foot a day, giant kelp keeps its fronds directed towards the sunlit surface with floats containing carbon monoxide.

FIVE

The California Coast

The small (one inch) dendronotid nudibranch in the upper lefthand corner is crawling on the holdfast of a kelp covered with white sponge.

The West Coast of the United States is radically different than either the East Coast or the Gulf of Mexico. The entire west coast of the Americas forms the west margin of the American Plate, consisting of North and South America and the western half of the Atlantic Ocean basin. The northern portion of the United States includes the coasts of Washington, Oregon and northern California. This part of the North American coast abuts a small piece of Pacific Ocean sea floor called the Juan de Fuca Plate, which for millions of years has been slowly moving eastward, wedging itself underneath these states and creating the present leading edge coast along the northern portion of the West Coast. This type of shore is characterized by an uplifted or block-faulted coast of a continent, facing toward a subduction zone or collision boundary, as is found here where the Juan de Fuca and North American plates meet.

As in the Gulf, rivers have played a major role in determining the nature of the West Coast sea floor. The Columbia, Fraser and other rivers have transported sediment such as mud, sand and gravel from coastal mountain ranges and dumped them on the Juan de Fuca Plate, where they have accumulated to a depth of over 2 miles. The sheer weight of the younger, top sediments forces the deeper, older particles to solidify into sedimentary rock. As in all other ocean basins, the platform upon which the sediments lie is solid basalt, whose greater density forces the Juan de Fuca Plate underneath the lighter continental rock of the northwest coast. As the Juan de Fuca Plate moves slowly eastward, however, the continental margin scrapes off most of the sedimentary rock, and the motion of the plate plasters this loose rock against the underside of the continent. As this wedge of plastered rock thickens over time, it gradually forces the edge of the continent upwards, creating a range of coastal mountains that now includes the Olympics, Willapa Hills, Oregon Coast Range and Klamath Mountains. The Juan de Fuca Plate, meanwhile, continues to descend beneath the continent at an angle of about 45 degrees, dragging down some of the sedimentary rock with it. When the sedimentary rock reaches a depth of about 30 to 60 miles, it melts under extreme heat and pressure, loses density and rises through the soft areas in the denser continental rocks. At the surface it erupts through volcanoes as basalt lava. On the West Coast these volcanoes form a north-south chain as part of the Cascade Range, located about 100 miles inland and parallel to the coast.

The West Coast is a leading edge

CHANNEL ISLANDS

Habitats
Kelp forests
Rocky shores
Sandy beaches
Seagrass meadows
Pelagic, open ocean
Deep rocky reefs

Key Species
California sea lion
Elephant and harbor seals
Blue and gray whales
Dolphins
Blue shark
Brown pelican
Western gull
Abalone
Garibaldi
Rockfish

Description
The sanctuary encompasses the waters surrounding San Miguel, Santa Rosa, Santa Cruz, Anacapa and the Santa Barbara Islands. A fertile combination of warm and cool currents results in a great variety of plants and animals, including: large nearshore forests of giant kelp, flourishing populations of fish and invertebrates and abundant and diverse populations of cetaceans, pinnipeds and marine birds.

Cultural Resources
Chumash Indian artifacts, shipwrecks

Location
About 25 miles off the coast of Santa Barbara, California

Protected Area
1,658 square miles

coast, and as such it is characterized by uplift, volcanic activity and the relative youth of its rock formations. These geologic activities express themselves as coastal features that include a range of coastal mountains, eroded sea cliffs, a steep gradient from coast range summits to the ocean, a narrow or nonexistent coastal plain, small bays and estuaries, a minimum of sand dunes and a shallow layer of continental shelf sediments.

In contrast, the East Coast is considered a trailing edge coast, characterized by a gently sloping shore, with barrier islands and a broad continental shelf, situated on the trailing edge of a continent and facing toward a spreading zone along a mid-oceanic rise. About 200 million years ago, North America separated from Africa and South America, creating the North Atlantic Ocean. The newly formed Atlantic Ocean defined a new shoreline along the margins of the separating continents. In the east, the North American continent and Atlantic sea floor are firmly attached and move together as a unit. There is no meeting of two neighboring plates that disfigures the coastal landscape by grinding, scraping and uplifting, and no spectacular demonstrations of volcanism such as the 1980 eruption of Mount St. Helens.

Kelp holdfasts provide food and shelter for a variety of organisms. Here a colony of brittlestars thrust their sinuose, writhing arms up and out of the holdfast haptera.

The most important geologic process at work on a trailing edge coast is erosion. Over a span of 200 million years wind and rain have eroded the east coast of North America, carrying sediments down rivers and into the Atlantic where they have created broad beaches, sand spits and barrier islands. So enormous is this delivery of sediment from the land that it has created a layer of sediment on the Atlantic shelf extending 180 miles seaward at a depth of up to 6 miles. Inland, this broad coastal plain extends 200 miles to the Appalachians, and has allowed the formation of such prominent embayments as Chesapeake and Delaware bays.

The two coasts are opposite not only in their tectonic and geologic history, but also in the development of the marine habitats found in each. The older and less dynamic trailing edge East Coast is dominated by broad intertidal flats, salt marshes and eelgrass beds. Burrowing organisms such as clams, polychaete worms and amphipods are well-suited to this kind of soft-bottom environment. Beach sands blow inland, forming sand dunes that move across the flat lowlands merging with the shore.

This southern kelp crab manages to hang on to slippery kelp fronds and stems with eight exceptionally sharp, pointed legs.

In contrast, the steep, uplifted shoreline of the leading edge West Coast does not encourage the formation of large estuaries and dunes, although some do exist, such as Padilla Bay in Washington and the extensive dune formations near Florence, Oregon. Instead, much of the West Coast is dominated by hard, exposed, rocky headlands creating tidepools and sea stacks that provide a place of attachment for mussels, barnacles and a variety of algae. These in turn provide food for sea stars, crabs, fishes, sea birds and marine mammals. While these kinds of marine environments can also be found in the East (particularly the Northeast), they dominate the intertidal coastal landscape in the West.

At the southern California coast the American Plate meets the Pacific Plate side by side. As the two plate margins slide past each other (the Pacific Plate is creeping northward at about three inches a year) they create a zone of plate contact that is not a single sharp rift or uplift. Instead a fracture zone has formed made up of many odd-angled faults of various sizes, the most famous (or infamous) being the San Andreas Fault. The results of a sliding edge coast are an assemblage of geologic formations of bizarre juxtapositions of rocks and soil types, creating irregular landforms and frequent earthquakes. San Francisco Bay, Tomales Bay, the Baja California peninsula and the Gulf of California were all created as the Pacific Plate slid along the margin of the American Plate.

Geology and Geography

The California çoast is about 1,798 miles long, including San Francisco Bay and its offshore islands. Some 70 percent of the coast is rocky cliffs or shores, although intermittent pocket beaches commonly punctuate the shore. The rest comprises either sandy fringing beaches backed by dunes, found south of San Luis Obispo, barrier beaches fronting lagoons and wetlands, found near Eureka, or the wetlands around San Francisco Bay. About 85 percent of the coast is actively eroding. Although long stretches of exposed west-facing shores remain largely natural, extensive stretches around San Francisco, Los Angeles and San Diego have been significantly altered by development.

Five distinct geomorphic provinces make up the California coast: the Klamath Mountains, Northern Coast Ranges, Southern Coast Ranges, Transverse Ranges and Peninsular Ranges. The geologic history of these five areas spans the past 225 million years of tectonic activity between the Pacific and North American plates.

Kelp's nemesis, purple sea urchins dine on kelp, often gnawing through the holdfasts and separating whole plants from them. The plants then drift away, taking with them large portions of the kelp forest.

CORDELL BANK

Habitats
Rocky subtidal
Pelagic, open ocean
Soft-sediment continental shelf and slope
Seamount

Key Species
Krill
Pacific salmon
Rockfish
Humpback whale
Blue whale
Dall's porpoise
Albatross
Shearwater

Description
Cordell Bank is an offshore seamount where the combination of oceanic conditions and undersea topography creates a highly productive marine environment. The bank rises to within 115 feet of the sea surface, with water depths of 6,000 feet only a few miles away. The prevailing California Current flows southward along the coast and the upwelling of nutrient rich, deep-ocean waters stimulates the growth of organisms at all levels of the marine food web. It is a destination feeding ground for many marine mammals and seabirds.

Location
60 miles northwest of San Francisco

Protected Area
526 square miles

California's primary geologic structural framework began to emerge approximately 130 million years ago, as a mountain-building period called the Nevadan Orogeny caused the ancestral Klamath Mountains, Sierra Nevada and Peninsular Ranges to rise from the shallow seas off-shore this early California coast. An island arc along the western margins of the North American Plate was also present at this time, and other geologic activity included general westward thrusting, volcanism and the build-up of enormous quantities of granite, which formed the underlying foundation of much of central California. The subduction of marine sedimentary and volcanic rocks in the deep trench along this plate margin, followed by uplift, produced the Franciscan Formation, a heterogeneous mix of easily erodible greywackes, shales and metamorphic rocks which exposes itself throughout these Coast Ranges.

The California state marine fish, garibaldi are members of the damselfish family, and characteristically create and defend specific territories from other members of their race.

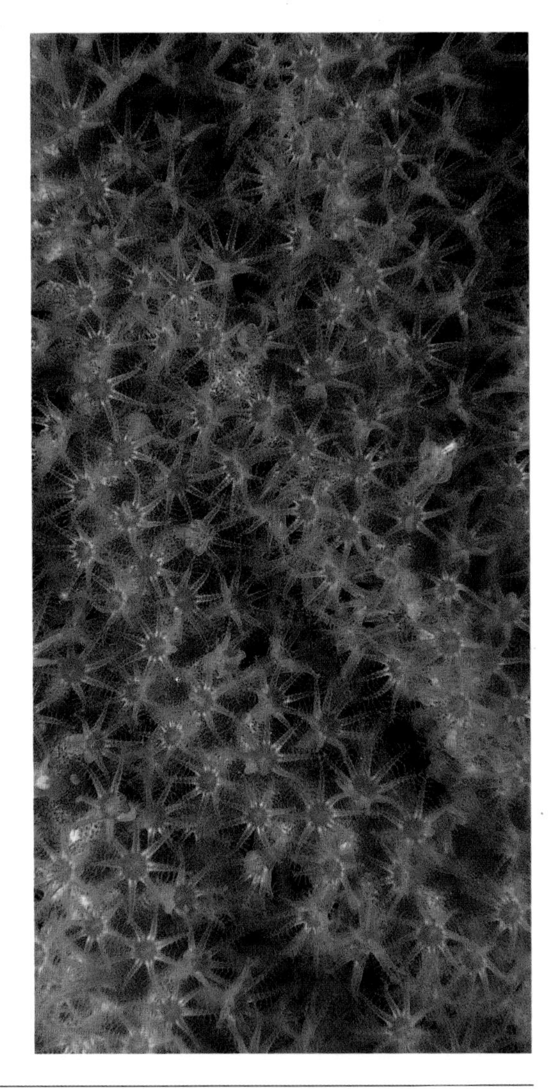

Soft corals exist in cold water, too. This close-up of a southern California sea whip (gorgonian) shows why they are also called octocorals—for the eight arms on each polyp.

Today's deformed and faulted marine terraces and subsiding coastal basins are evidence of the past 2 million years of active tectonism. Terraces are rock platforms that have been in place long enough to erode evenly by wave action. Tectonic uplift raises them above the action of the waves, halting erosion by the sea and preserving their general step-like appearance. A good example of terracing occurs on the Palos Verde Peninsula in Los Angeles, where a "flight" of 13 marine terraces has its highest "step" 1,140 feet above sea level. Fragments of high terraces are found throughout the coast, from a deformed sequence near Eureka along the seaward slopes of the Coast Ranges, to the extensive mesas around San Diego, where the 270- to 450-foot Linda Vista terrace and its fossil barrier beaches are considered 1 million years old. The presence and placement (in time and space) of these terraces is evidence that tectonic activity has been fairly continuous over the past several million years, and remains so today, as every Californian knows.

While the periodic rise of the California coast has occurred primarily through tectonic uplift, there is also evidence that at times this coast has become submerged as a result of changes in sea level and tectonic subsidence. For example, the Pacific Ocean entered the Golden Gate about 10,000 years ago and, rising at a rate of 2 centimeters per year, flooded laterally across San Francisco Bay as rapidly as 90 feet per year up to 8,000 years ago. The rate of sea-level rise then declined and for the past 6,000 years has averaged .04 to .08 inches per year. The actual placement of the California coast with respect to sea level must take into account tectonic uplift, subsidence and changes in sea level over time, and the fact that global sea level continues to rise, not lower.

GULF OF THE FARALLONS

Habitats
Coastal beaches
Rocky shores
Mud and tidal flats
Salt marsh
Esteros
Pelagic, open ocean
Deep benthos, continental shelf, slope

Key Species
Dungeness crab
Gray whale
Steller sea lion
Common murre
Ashy storm-petrel

Description
The sanctuary includes nurseries and spawning grounds for commercially valuable species, at least 26 species of marine mammals and 15 species of breeding seabirds. One quarter of California's harbor seals breed within the sanctuary. The Farallon Islands are home to the largest concentration of breeding seabirds in the continental United States. The sanctuary boundaries include the coastline up to mean high tide, protecting a number of accessible lagoons, estuaries, bays and beaches for the public.

Cultural Resources
Shipwrecks
Fossil beds

Location
Along the coast of California north and west of San Francisco

Protected Area
1,255 square miles

Much of the California coast is under attack by erosional forces, both from the land and the sea. Along the northern California coast, erodible rocks, steep slopes and high precipitation and runoff combine to produce frequent landslides and high erosion rates north of Point Conception. The mass movement of earth from sea cliffs and coastal slopes also contributes to the sediments that wind up on the West Coast continental shelf, notably where fractured or poorly consolidated rocks occur. This source of sediment has been both aggravated in some places and curtailed in others by peoples' activities, primarily through road and other construction, such as along the stretch of coast near Malibu west of Los Angeles and south along the Big Sur coast.

Purple hydrocoral is one of the most delicate members of California's rocky reef habitats. While it takes a variety of forms, it generally can be found as an intricate framework of fine branches, usually at depths below 70 feet.

Offshore Bottom Topography

Offshore California is represented north of Point Conception by a shelf and

Strawberry anemones are a dominant life form on the rocky reefs of the California coast. This extreme close-up (each polyp is a half-inch in diameter) clearly shows the characteristically clubbed tentacles.

slope, and southward by what is called the continental borderland. The relatively narrow northern shelf widens to 31 miles off the Golden Gate, where a large crescent-shaped sandbar composed of terrestrial sediments can be found. The continental/borderland off southern California comprises a series of fault blocks and troughs with some closed basins and rugged emergent islands. At least 32 distinct submarine canyons dissect California's offshore area, notably Delgada, Noyo and Bodega canyons off northern California and Sur, Lucia and Arguello canyons farther south. Their presence is indicative of a time when this shallow shelf area was exposed to the affects of surface erosion from rivers. A period of subsidence and a rise in sea level has left them as submerged features of the California coast.

Perhaps the most spectacular of these submarine canyons is found in Monterey Bay. Monterey Bay is approximately 25 miles in length, from Soquel Point on the north to Point Pinos on the south, and from 5 to 13 miles wide. The Monterey Canyon almost exactly bisects the bay. It rivals the Grand Canyon (one third as long; twice as deep) and looks similar to it in cross-section. The canyon slopes down from a depth of about 60 feet at its head at Moss Landing to over 12,000 feet at its end 60 miles off the coast. Most of the bay's sea floor is covered with soft-bottom sediments. The coarse sand near shore becomes finer in deeper water, a result of decreasing wave action, which otherwise washes away fine particles. Sandy beaches stretch from Santa Cruz to Monterey.

The West Coast has the highest diversity of sea stars anywhere in the world. Here a prickly star (right) may be chasing down a bat star. Many sea stars prey on each other.

Geologists believe that the Monterey Canyon was not formed at its present location, and that it has been filled with sediment and re-excavated several times in the last 30 million years. As the Pacific Plate moved north at about 3.9 inches per year over the last 20 to 30 million years, it brought the Monterey Canyon—and the rest of coastal California up to San Francisco—along the San Andreas Fault to its present location. Over time the canyon has been eroded in two ways: underwater, by sandfalls called turbidity currents, and when it was exposed to the air, which occurred when the continental margin was uplifted and sea level dropped during recent ice ages.

Another notable shelf feature includes a series of submerged banks and seamounts scattered along the central-southern California coast. One of the most well studied of these is the Cordell Bank, located approximately 25 miles off Point Reyes on the very edge of the continental shelf. The shallowest part of this "underwater island" is about 120 feet deep, while most of its top averages 180 feet deep. During periods of glaciation and lowered sea levels, this bank was actually exposed to the surface as seen by the presence of wave-cut terraces. These correspond to periods

Like its warm-water relatives, the California moray is also a night-active fish, hanging out in the shadowed cracks and crevices of rocky reefs during the day.

MONTEREY BAY

Habitats
Sandy beaches
Rocky shores
Kelp forests
Submarine canyon
Pelagic, open ocean
Wetlands

Key Species
Sea otter
Gray whale
Market squid
Brown pelican
Rockfish
Giant kelp

Description
Monterey Bay, the nation's largest marine sanctuary, spans over 5,300 square miles of coastal waters off central California. Within its boundaries are a rich array of habitats, from rugged rocky shores and lush kelp forests to one of the deepest underwater canyons on the West Coast. These habitats abound with life, from tiny plants to huge blue whales. With its great diversity of habitats and life, the sanctuary is a national focus for marine research and education programs.

Cultural Resources
Indian midden sites
Naval airship USS *Macon*

Location
Central California coast

Protected Area
5,328 square miles

of glaciation and changes in sea level over time. It is now part of the National Marine Sanctuaries Program, as is Monterey Bay.

Oceanography

The oceanography of the West Coast is directly tied to offshore weather patterns and the way high and low pressure systems affect the movement of water throughout the eastern Pacific. Further, the nature of West Coast oceanography is as different from the East Coast as it is tectonically. For example, on the eastern seaboard air temperature regulates water temperature for much of the year, primarily due to the vast extent of its shallow continental shelf. On the west coast the opposite is true. Here the continental shelf is narrow in most places, and water temperature plays a significant role in regulating coastal weather patterns. The extent of this control ranges from northern California to Washington. At Point Conception things change dramatically, both in terms of the physical nature of the coast as well as the abundance and kinds of marine species found north and south of this headland.

Banded gobies, like the one peering out from the hole in this Astrea shell, are exceptionally common on southern California rocky reefs south of Point Conception.

The California Current

The primary current affecting water motion along most of the West Coast is the California Current. It extends 20 to 150 miles from shore, and moves at the sluggish pace of about 4 to 8 miles a day. Consisting of sub-Arctic water, the California Current is characteristically cool, low in salinity and high in nutrients.

It is the southern branch of the North Pacific Drift, which originates at about 40° latitude as a result of the prevailing westerlies producing a more-or-less westerly ocean current. In the Gulf of Alaska it merges with the Subarctic Current, which travels at about 1 to 2 miles per day. The Subarctic Current reaches shore near Vancouver Island and in the summer breaks into the southward-flowing California Current and the northward Alaska Current. The northerly winds on the east hub of the North Pacific High drag the California Current south and cyclonic storms maintain the Alaska Current at about 8 miles per day.

A stout fish, the monkeyface eel grows to about two feet long and lives among the seaweeds, kelps and eelgrass beds of shallow, rocky coasts along the entire Pacific coast.

Seasonality plays an important role in the extent and force of the California Current as it changes significantly between summer and winter months.

In the summer, a strong clockwise outflow of air derives from the Hawaiian high-pressure cell to the west, and prevailing winds reach the coast from the northwest. In the winter this cell is weaker and farther south, but counterclockwise flow toward the Aleutian low and eastward progressions of cold fronts across California maintain these prevailing winds for much of the time. North of Point Conception, 30 to 50 percent of ocean swells approach from the northwest and most others from the west-northwest or west. The predominantly northwest swells set up strong longshore currents, although in some places the opposite may occur—a northward drift may take over as a result of shore configuration, bottom topography and reversing currents and eddies.

An important consequence of the California Current is the phenomenon of upwelling, whose repercussions are felt all along the West Coast. As wind blows across the Pacific, frictional forces "pull" a mass of water at the surface. Because of the earth's rotation, the currents do not follow the winds exactly. This is called the "Coriolis effect" and it is the concept that relative to the earth, any moving object veers to the right in the Northern Hemisphere and to the left in the Southern, and the veering increases with latitude. As a result, any surface ocean current in the Northern Hemisphere travels at an angle 45 degrees to the right of the wind direction.

Sleek, fast-swimming blue sharks frequent the open ocean waters off the coast of southern California. They are among the most abundant in the world.

Because of the Coriolis effect, the north winds of summer produce a surface current along the West Coast that tends to become somewhat directed to the west. As this surface water moves offshore, it creates a lower pressure or space along the shoreline, and deeper water moves up to take its place. Water from as deep as 600 feet moves upward at speeds reaching 60 feet per day. Relative to the surface, the deep water, much of it originating from northern Arctic waters, is cold, high in salinity and nutrients and low in oxygen. This water emerges at the surface in a narrow band that hugs the shore from the surf zone out to about 10 miles, where it begins to mix with the surrounding sea.

Upwelling is most pronounced where the north winds are strongest—along Washington, Oregon, the central California coast and around headlands, which themselves tend to divert currents offshore. On the northwest coast, upwelling seems to be most intense between Capes Blanco and Mendocino, off the central Oregon coast from Newport to Florence and off Tillamook. At these locations temperatures in the surf zone average 43 to 45°F in the summer while some 50 miles offshore, in the surf between Capes Arago and Blanco, and along most of the Washington coast they average a relatively warm 55°F. This upwelled water often cools the overlying humid air enough to precipitate fog. This phenomenon is particularly evident at San Francisco, and a good example of how Pacific Ocean oceanography controls West Coast weather patterns. As warm, moisture-laden air moves across the Pacific, it encounters the cold surface waters of the California Current and nearshore upwelling, and is suddenly cooled. This cooling causes the water vapor to precipitate in the form of fog, which occurs commonly along this coast in the summer.

Disturbed from its sandy hiding place, a large bat ray lifts off from the bottom in southern California's Catalina Island.

The summer rise in nutrients has significant ramifications for the ecology and energy budget of West Coast marine life. The dramatic inflow of available nutrients, combined with the high light levels found at the surface, cause an explosion of phytoplankton populations so significant they are called "blooms." These concentrations of microscopic plants provide food for millions of zooplankton (mostly copepods) which in turn experience their own population explosions. The zooplankton provide another source of food for filter-feeding invertebrates and vertebrates, extending this sudden availability of

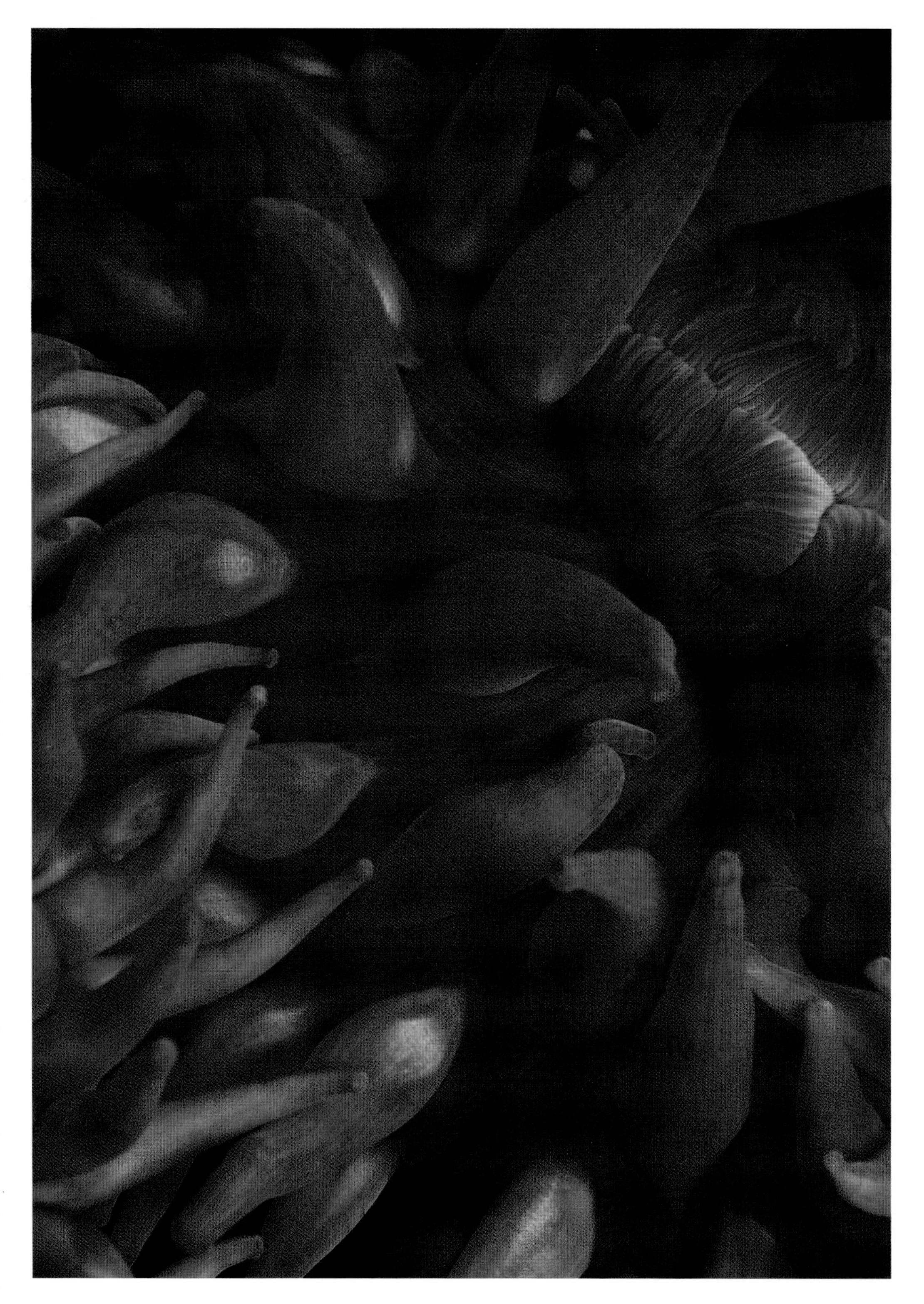

organic energy all the way up the food chain to seabirds, marine mammals and even humans. Because of upwelling, phytoplankton productivity in spring and summer is considerably higher on average along the nearshore than offshore, and in the fall and winter it drops to levels comparable to offshore.

When the cyclones move to the south during winter, their powerful southwest winds reverse the California Current off Washington, Oregon and northern California and a new current takes affect—the Davidson Current. The Davidson Current is a countercurrent, flowing north between the southerly California Current and shore. This countercurrent flows from

This close-up of a giant anemone's tentacles reveals their symmetrical, repeating pattern of distribution around the central mouth.

Baja California up beyond Mendocino. Though it usually moves beneath the surface at a depth of about 600 feet, it surfaces from December through February. Davidson Current water is characteristically warm (52°-54°F) and high in salinity. During this time nearshore winter temperatures average 47°F off northern Oregon and Washington, 49°F off central Oregon and between Capes Blanco and Mendocino are actually warmer than in the summer (51°F). The Davidson Current travels northward at about 12 to 24 miles per day and buffers the region somewhat from cold temperatures. Occurring from November through February, this period is also characterized by the lowest salinities

Giant anemones visually punctuate the subtidal rocky reef habitats of the Pacific coast from Alaska to southern California. Up to a foot in diameter, they are capable of capturing small fish with their stinging tentacles.

of the year, slowly cooling surface temperatures and little change in temperature from the surface down to 160 feet.

South of Point Conception the central California shoreline takes an abrupt curve east, forming the Southern California Bight, and water temperatures rise in general. The dramatic change in coastal topography and orientation mitigates the affect of the California Current and its cooler waters. Also, the presence of the Channel Islands gives the inshore waters to the south of the point some additional protection from the southbound California Current. These changes in topography, and onshore current movement from the south during the winter, allow relatively warm southern water to creep into the area along

A collection of tiny amphipods ("sea fleas") can be seen here scavenging for bits of undigested food around the mouth of a giant anemone.

TIJUANA RIVER

Habitats
Dunes and beaches
Mudflats
Salt marshes
Riparian
Coastal sage
Upland

Key Species
Light-footed clapper rail
California least tern
Least Bell's vireo
Salt marsh bird's beak
Cordgrass

Description
Tijuana River is an intertidal coastal estuary on the international border between California and Mexico with three-quarters of its 1,735-square-mile watershed in Mexico. The salt marsh dominated habitat is characterized by extremely variable streamflow, with extended periods of drought interrupted by heavy floods during wet years.

Location
San Diego County, Imperial Beach, California

the southern California coast. Here 70 percent of swells pass through the Santa Barbara Channel from due west, while 80 percent of swells approach Los Angeles from the west southwest. Warm-water movement from the south is also encouraged by late summer hurricanes off western Mexico, Southern Hemisphere winter storms and local winter depressions passing along more southerly tracks. Water of similar temperature extends south all the way to the temperate-tropical boundary at Magdalena Bay, Baja California. There are, however, a few spotty interruptions caused by local upwelling.

Zoogeography: The Southern California Region

Without a doubt, the most important physical coastal feature influencing the distribution of marine life along the California coast is Point Conception. Just as Cape Hatteras represents a major break in the distribution of East Coast fauna, so does Point Conception on the West Coast. It is here that the influence of a warm-temperate water regime flowing into the Southern California Bight meets the colder waters flowing south via the California Current. This physical and temperature barrier plays a significant role in the distribution of marine life all along the West Coast.

South of Point Conception, California can be divided into two zoogeographic provinces: Cortez and San Diego. The Cortez province takes in the area south of the Mexican border, including all of the Gulf of California (also known as the Sea of Cortez). The San Diego province is the area south of Point Conception to the border. These two provinces of the California Warm-Temperate Region are quite different from one another. The endemic fauna of the Gulf of California has been almost entirely derived from the Eastern Pacific

ELKHORN SLOUGH

Habitats
Oak woodland
Salt marsh
Grassland
Mudflat
Freshwater pond
Monterey pine grove
Coastal scrub

Key Species
Shorebirds
Great blue heron
Great egret
Smoothhound and leopard shark
Birds of prey

Description
Elkhorn Slough features mixed oak woodlands and grasslands on rolling hills overlooking tidal salt marsh. Groves of eucalyptus and Monterey pine and several small freshwater ponds dot the reserve. Habitat enhancement and restoration projects are presently being initiated.

Location
Extends 7 miles inland east of Moss Landing, California

Tropical Region to the south, while that of the San Diego province demonstrates a dual origin, about half being related to northern families and genera and about half to tropical groups. In the Gulf, with the exception of a small group of species that are shared with the San Diego fauna, virtually all the nonendemics may be classified as tropical species. In contrast, about two-thirds of the nonendemics along the outer coast section are temperate species that range into the area from north of Point Conception, the remaining third having tropical affinities.

The Cortez Province

The geographic position of the Gulf of California is of interest to marine zoogeographers because its mouth is in the tropics while the remainder of the Gulf is situated far enough north to be considered warm-temperate. This condition has apparently prevailed at least from the beginning of the Pleistocene (approximately 2 million years ago) since, by then, the peninsula of Baja California had been formed to approximately its present outline and extent.

Apparently only tropical organisms had open access to the Gulf of California during the Pleistocene. The entrance to this body of water is relatively wide and deep, and for the most part Pacific Ocean water can flow freely in and out of the Gulf, resulting in little, if any, restrictions to the dispersal of most marine organisms. However, it would appear that the primary reason for the development of a highly endemic fauna and flora in the Gulf is the presence of a water temperature barrier, rather than a physical one. Over the past 3 million years (the approximate duration of the Pleistocene) a distinct warm-temperate fauna has developed from the tropical one which originally evolved in this area. The presence of a temperature barrier alone will produce sufficient isolation to permit this process to take place over such a period of time. The Gulf of California is considered to be a distinct warm-temperate province. It is wholly isolated from the northern San Diego province, with which it has relatively few species in common. Its many endemic species were independently derived from tropical forms to the south, and the rest of the fauna (with the exception of some warm-temperate invaders from the north) is comprised of eurythermic tropical species.

The San Diego Province

The marine flora and fauna of the California coast south of Point Conception is influenced primarily by the rich, cold-temperate area of the North Pacific, with additional input of species coming from the southern Cortez province. For example, 130 (45 percent) out of 219 species of fishes on the northwestern coast of Baja entered that area from north of Point Conception. In comparison, only 65 were tropical species of southern origin and 96 were provincial endemics.

In southern California, the northern species play an even larger role as evidenced by the appearance of additional wide-ranging, cold-temperate species and a decrease of tropical forms. This area is an interesting zone of mixing, where both southern and northern eurythermic species are brought into contact. At the same time, about 30 percent of the marine plants and animals are endemic species. More northern species are found on the deeper portions of the continental shelf, apparently due to a distributional principle called "isothermic submergence." This is where animals of wide latitudinal distribution occur in deeper water as one moves south of their origin. Species such as the sea anemone, *Metridium senile,* which appear abundantly in deep water off California, can be found in shallower waters and tidepools as one moves north.

Pacific Coast Algae

One of the prime beneficiaries of the nutrient-rich upwelling that comes with the California Current are the large algae that line the rocky shores of the entire West Coast. While these species can be found in the higher latitudes of the eastern seaboard (primarily those under the influence of the cold Labrador Current), their colonization of the shallow subtidal in the East is not nearly as extensive as in the West. Many of these large marine plants have a truly amazing range. Giant kelp, for instance, can be found from southern California to southeast Alaska. The presence of these plants is also one of the reasons why west coast fauna is generally richer and more diverse than on the east coast. They provide a complex habitat housing many small invertebrates and fishes, in addition to providing a ready food source for many of the same. The surface of almost any large seaweed is covered with a myriad of worms, hydroids, bryozoans and other animals living on the seaweed and in its holdfast.

Seaweeds stand in stark contrast to the primary producers of the sea, the microscopic plants (phytoplankton) that float freely in the ocean. There are at least 6,000 known species of phytoplankton, and they account for most of the ocean's vegetation, producing a significant amount of the Earth's oxygen. Seaweeds, on the other hand, are large, many-celled marine algae, the most prominent plants on open coast rocky shores and subtidal reefs. Most algae belong to one of three groups: green, brown and red. Most green algae (*Chlorophyta*) are distinctly green; most brown algae (*Phaeophyta*) are brownish, including the large kelps and most red algaes (*Rhodophyta*) are reddish, but this group varies the most in color—they may also be purple, yellow, green or brown. California's marine algae are among the richest and most diverse in the world; more than 280 genera and 660 species have been identified to date. Algae can be found in nearly every shallow subtidal and intertidal habitat, ranging from protected estuaries to rocky headlands exposed to the full force of the sea.

Kelp

Perhaps the most important algae living on the West Coast is the giant kelp, *Macrocystis pyrifera*. Forests of these giants are complex natural communities that provide food and shelter for a variety of other organisms. This brown algae grows at a rate of a foot a day and produces individual plants 60 to 80 feet tall. "Groves" of these plants create a unique subtidal habitat—kelp forests—which provide a complex habitat for thousands of other marine plants and animals.

A giant kelp plant has several distinct parts. Haptera, a cone-shaped holdfast made up of root-like branches, attaches the plant to rocks or other solid substrates. A number of fronds grow from the top of the holdfast. Each frond includes a rope-like stipe and numerous blades, each with a gas-filled float at its base. The floats support the fronds, directing them toward the surface, and the light. New blades are added at the tip of the frond and the fronds get longer as the stipe grows between the blades. During the peak season, fronds may grow 10 inches per day. Once they reach the surface, the fronds continue to grow, forming a floating canopy. As old fronds die, new fronds sprout and grow from the top of the holdfast. Specialized blades called sporophylls grow at the base of each frond. These blades, which lack floats, are used for reproduction.

F I V E

From Alaska to Baja, giant kelp forms underwater "forests," creating a vertical landscape unique to the aquatic realm. These huge plants, up to 100 feet tall, provide a complex habitat for thousands of other species.

In the Southern Hemisphere, giant kelp grows along nearly all temperate coasts where upwelling occurs. In the Northern Hemisphere giant kelp grows only along the West Coast of North America, from northern Baja California to central California. A few isolated populations can be found from central California up to Alaska, but generally speaking, the bull kelp (*Nereocyctis leutkeana*) is the dominant kelp forming forests further north. Giant kelp forests grow only in cool, nutrient-rich waters along temperate coasts below 70°F. It is not known whether this is due solely to temperature or a more complex interaction of temperature and nutrient levels, since as temperature rises the amount of nutrients dissolved in sea water decreases. Other factors that influence where giant kelp grows include the type of surface available, depth, availability of light, turbidity and water motion. Kelp typically grows on rock outcrops, and evidence of a kelp bed from the surface almost always means there is a rocky reef below.

The relationship between water depth and the penetration of sunlight is particularly important to the distribution of kelp as it relates to the plants' ability to photosynthesize. Kelp cannot grow at depths where there is less than 1 percent of the light present at the surface. This depth varies between 10 and 100 feet, depending upon water clarity. The depth at which kelp grows can also be affected by the presence and thickness of the surface canopy created by the topmost fronds of the plants laying flat on the surface. Obviously, the thicker the canopy, the more shading takes place, decreasing the amount of light striking the entire plant.

Water motion is also important to kelp. Steady, gentle water flow seems to aid the plants in taking up nutrients from the surrounding sea water. However, during winter storms whole kelp plants may be ripped off the bottom. Once a plant is torn loose, its fronds often become entangled with neighboring plants, increasing the surface area upon which the waves may act and pulling them up, too. When they become unattached, kelp plants leave the forest as islands of "drift algae." Far out to sea, these masses of seaweed continue to provide habitats for a variety of juvenile and adult fishes. Consistently strong waves can bury young plants with constantly shifting sand or rip out larger plants. At other sites, kelp forests may appear during the summer and be removed the following winter. Bull kelp is not as easily broken and often occurs in areas too rough for giant kelp.

The Kelp Forest

Anyone who has swum through a kelp forest on a bright sunny day will attest to its similarity with land-based forests. As in a forest on land, the presence of kelp plants changes the physical environment. Kelp is the only natural physical feature that adds a vertical element to the aquatic realm. A dense stand of kelp will decrease water motion caused by swells, creating a zone of calm, protective water. This space becomes home for many juvenile fishes, including rockfishes, senorita, kelp and surf perch, who spend the early part of their lives in kelp forests, feeding on plankton concentrated in this "quiet zone" in the ocean. The presence of giant kelp also affects other plants living in and around a kelp forest. Since the surface canopy reduces the amount of light that reaches the bottom, this may affect the type and distribution of other plants living nearby. On land, the forest is not made up of only one species of plant, but an assemblage of trees, bushes, grasses and other plants that create the physical structure of the forest. The same is true underwater in a kelp forest.

In addition to creating the general structure of the forest, kelp and other large marine plants provide a variety of microhabitats for other smaller plants, invertebrates and fish. The canopy, the fronds and the holdfast all provide places for hundreds of different organisms to live. Sea otters and shorebirds use the canopy for shelter or places to forage. Encrusting organisms such as hydroids, bryozoans and tube worms actually utilize kelp and other large plants for places of attachment. Turban snails, crabs and isopods all live on the blades and stipes. Shrimp-like mysids and small fishes like the senorita and surf perch live in the water between the plants, closely associated with the fronds, using them for shelter and hiding places. At the base of the kelp, the holdfast provides a complex three-dimensional space similar to a coral reef structure. This part of the forest is home to thousands of amphipods, brittlestars, crabs and small invertebrates that live among the tangled holdfast.

Besides creating its own environment, the plants of the kelp forest, particularly giant kelp, are a major source of food for herbivorous invertebrates and fishes. The rapid growth of kelp means that large amounts of organic material are being created through photosynthesis. This food energy is passed on to marine animals that consume kelp directly. Turban snails, kelp crabs and isopods, as well as fishes like the halfmoon and opaleye feed directly on kelp. In turn, these organisms become food for predators such as starfish, crabs, other fishes and even mammals like seals and sea otters, which eat many of these herbivores.

Detached pieces of kelp provide food for animals who do not even live in the kelp forest proper. Called "drift algae," this organic plant material is a major source of nutrition for many marine animals. Over half of all the drift produced by a forest winds up in other marine communities as a result of storms tearing away whole plants or the simple decay of a dying plant. Many organisms living in places where kelp does not grow, such as deep reefs and sandy sea floors, depend on drift kelp for much of their food. The export of kelp extends even to land communities when kelp parts wash up on a beach. Here it becomes food for kelp flies, beach hoppers and other animals that roam the intertidal.

When fronds break off and sink to the bottom of the forest, abalones, sea urchins, bat stars and others consume the detached drift algae. Eventually, most drift algae that remains in the forest is broken down by bacteria, creating a fine detritus that provides food for filter-feeders such as barnacles, tubeworms and sea cucumbers.

Kelp Forest Structure

No two California kelp forests are the same. They vary in many ways, from the extent of area they cover to the kinds of algae forming the forest. However, there are some aspects of kelp forest ecology and structure that hold true for most kelp forests. As with land plant forests, there is a clear vertical stratification of plants and animals within most kelp forests, which can generally be divided into four layers. The surface, or primary canopy, is usually composed of giant kelp or bull kelp. Below the surface canopy is the understory, or secondary canopy, which is made up of a variety of smaller brown algae such as *Laminaria,* and *Pterygophora.* These plants are usually less than six feet tall. Underneath the secondary canopy are a variety of small, low-lying red and brown algae such as *Gigartina, Corallina, Rhodymenia, Dictyoneurum* and *Desmarestia.* These species are generally much less than three feet tall. The fourth layer is the closest to the substrate; it is the crustose coralline layer. Coralline algae comes in two forms: a thin, sheet-like form that grows on rocks looking something like a lichen and a more crustose form that has articulating branches. Both incorporate calcium into their "bodies," partially as a defense mechanism against browsing herbivores.

Each forest layer provides food and shelter for a specific group of animals. There are also many who roam throughout all four layers, particularly fishes and more mobile creatures.

The development and species composition of the different layers also vary with the specific environmental conditions imposed by the area in which it is found. Kelp forests in semi-protected, calm areas have sparse understory, while those in exposed, wave-swept areas may have no surface canopy and a thick understory.

Seasonal changes and irregular, unpredictable disturbances also cause variability. For instance, if winter storms remove the primary and secondary canopies, light levels on the bottom increase and the remaining low-lying plants may grow into a lush, thick cover. Only when this bottom cover is thinned or removed can the larger species recolonize and grow. Once they do become re-established, they persist by shading out other plants. Again, as on land, light penetration through the forest is a key concern for plants in the understory and on the forest "floor."

Water depth is also a key factor in determining the species composition and structure of a kelp forest. Where a species grows may depend on differences between water motion, light, competition with other algae and grazing at different depths. For instance, shallow-water species like the feather boa kelp *Egregia* are generally more tolerant of rough water and high surf conditions. Because large canopy-forming species are massive they would easily be ripped up in such rough water motion, so they grow in deeper, calmer water. In deeper water still, few algaes grow at all, probably because light levels are too low to allow for photosynthesis. Here, stationary invertebrates reign, including bryozoans, sponges, cup corals and colonial tunicates, who all compete for a hard place to attach on the same rocks to which the giant kelp is attached. There are also differences in the assemblage of plants and animals that attach to horizontal and vertical rock surfaces. Plants dominate horizontal surfaces, while animals dominate most vertical surfaces. This may be due to differences in light levels, sedimentation accumulation, the differences in patterns of water flow or the preferences of invertebrate larvae and plant spores.

Most changes in kelp forests occur in cycles. For instance, the canopy thins out in winter and grows back in summer. Even unusually severe disturbances such as winter storms are usually followed by a predictable period of recovery. However, over time other changes have occurred in kelp forests along the California coast that have had wide-sweeping effects on the overall population of kelp forests. Between 1940 and 1960, many of the southern California kelp forests deteriorated drastically. The main cause of this loss was intense grazing by large numbers of sea urchins. Sea urchins generally feed on drift algae. But when their numbers increase they begin to feed on living and attached kelp plants as well. The weakened fronds break easily and drift away. Even though the sea urchins actually eat very little of the kelp, they can quickly clear large areas, leaving only holdfast stumps behind. Like beavers cutting down trees, large numbers of urchins can effectively clear-cut a kelp

forest, removing the entire habitat and many of the other plants and animals associated with it.

Otters, Urchins and Kelp

It is still unclear why the urchin population increased during this time period. One theory, however, implicates the sea otter, *Enhydra lutris*. Sea otters are eating machines that consume up to 25 percent of their body weight each day (an adult female weighs 40 to 50 pounds). Typical prey includes abalones, sea urchins, crabs and other invertebrates and the occasional slow fish. They once ranged from the Aleutian Islands to Baja California, but fur traders hunted them to near extinction along the California coast over 170 years ago. Because of their need to eat large quantities of food (related to heat loss and the intense time they spend grooming their fur) sea otters have a tremendous impact on the subtidal communities in which they live. Where there are otters there is lots of algae but few invertebrate grazers like sea urchins. Conversely, in most areas where there are no otters, there are more invertebrate herbivores and less algae.

This California sea otter will spend most of its life at sea near kelp forests. Thick, dense fur keeps the otter warm in the cold ocean water.

One theory that has been proposed to explain these observations is that when hunters removed sea otters from southern California, sea urchin populations increased, upsetting the checks and balances of the system, which then destroyed the kelp forests. Although this explanation seems to make sense superficially, it is probably incorrect. Even though sea otters can have important effects on giant kelp populations, their absence has had little to do with fluctuations of southern California kelp forests over the last four decades. Sewage, pollution, unusually warm water temperatures and possibly human activities (removing sea urchin predators like spiny lobsters or competitors like abalones) have probably been more influential.

Sea otters disappeared from southern California over 150 years before the decline of the kelp forests began. If the sea otter had been the major factor controlling sea urchin populations, they would have probably increased before 1940. Second, pollution was high in some areas where the loss of kelp was the greatest, notably large sewer outfalls in Los Angeles and San Diego. Raw sewage from these outfalls increased turbidity of the water and covered the substrate with sediment and sludge. Lower light levels and sedimentation would probably have damaged kelp populations even if there had been no overgrazing by sea urchins. Lastly, several successive years of unusually warm water occurred in the late 1950s that were probably the final blow for the kelp populations. The combination of pollution, warmer water and an increase in the urchin population all came together at the same time to create the conditions that destroyed the kelp beds in the '40s, '50s and '60s.

The role sea otters play in this dynamic, relatively sensitive system is still somewhat unclear. In recent years sea otters have been re-introduced to parts of the central California coast from Alaska, and their populations have increased, expanding their range. Kelp forests have improved in some of these areas. However, this does not mean that the kelp forests would be destroyed by sea urchins if there were no otters. Studies of kelp forests where there are no otters, such as the exposed coast of north Santa Cruz, have shown that severe winter storms remove all urchins not protected by cracks and crevices. In this scenario the weather controls the populations of urchins, and the resulting health of the kelp forest.

Other studies in southern California have found that a region can be abundant in both sea urchins and kelp, and illustrate the interdependence of all the organisms within the kelp forest community. Kelp forest areas with rocky reefs having high vertical relief generally have more kelp since they provide shelter for predators like sheephead and spiny lobsters that eat sea urchins. "Mountainous" underwater terrain also traps more drift algae, providing more food for urchins and other herbivores that does not affect the health of the living plants. Flat, rocky reef areas tend to support kelp forests that fluctuate more. All kelp plants may be removed from these areas, which are then recolonized from nearby forests until the urchins remove them again. The ongoing health of kelp forests is a complex balance of physical factors such as winter storms and water temperature and biological factors such as the number of spiny lobsters available to feed on urchins.

This eelgrass jellyfish doesn't act like a jelly at all. The points on its tentacles that appear to be bent indicate the location of small suction pads which allow this jelly to attach to the sea floor, usually to the blades of eelgrass.

SIX

The Pacific Northwest & Alaska

There are approximately 50 species of rockfish on the West Coast, and one of the most common is the China rockfish. These big-eyed fish can be seen hovering over rocky reefs and in kelp beds, calmly surveying the underwater scenery.

Marine life along the West Coast is very similar wherever you go. Large algae grow in most areas where there is a solid surface and many species of fishes and invertebrates can be found along the entire western margin of the United States and into southeast Alaska. The aptly named "ubiquitous" nudibranch—*Hermissenda crassicornis*—can be found from the Channel Islands up into Alaska. The plumose anemone, *Metridium senile*, is another hallmark West Coast species that also inhabits the colder parts of the East Coast. It is one of the few bi-coastal organisms of the country.

There are both dramatic and subtle changes in the characteristics of north Pacific marine life. As already mentioned, some cold-loving species, *Metridium*, for

example, are found in shallower water farther north. Other species disappear altogether north of Point Conception. The main physical factor influencing the distribution of marine life along the West Coast is water temperature, and this varies regionally as a result of current patterns, tides and season.

Continental Shelf and Slope

The continental shelf off the coast of northern California, Oregon and Washington is relatively narrow, especially when compared to the broad shelf on the East Coast of the United States. Oregon's continental shelf is a flat, gently sloping

Red Irish lord on the half shell, Puget Sound, Washington.

The longfin sculpin is a small fish, generally not more than a few inches long. This individual is nestled into one of its favorite hiding places, a bed of colonial sea squirts.

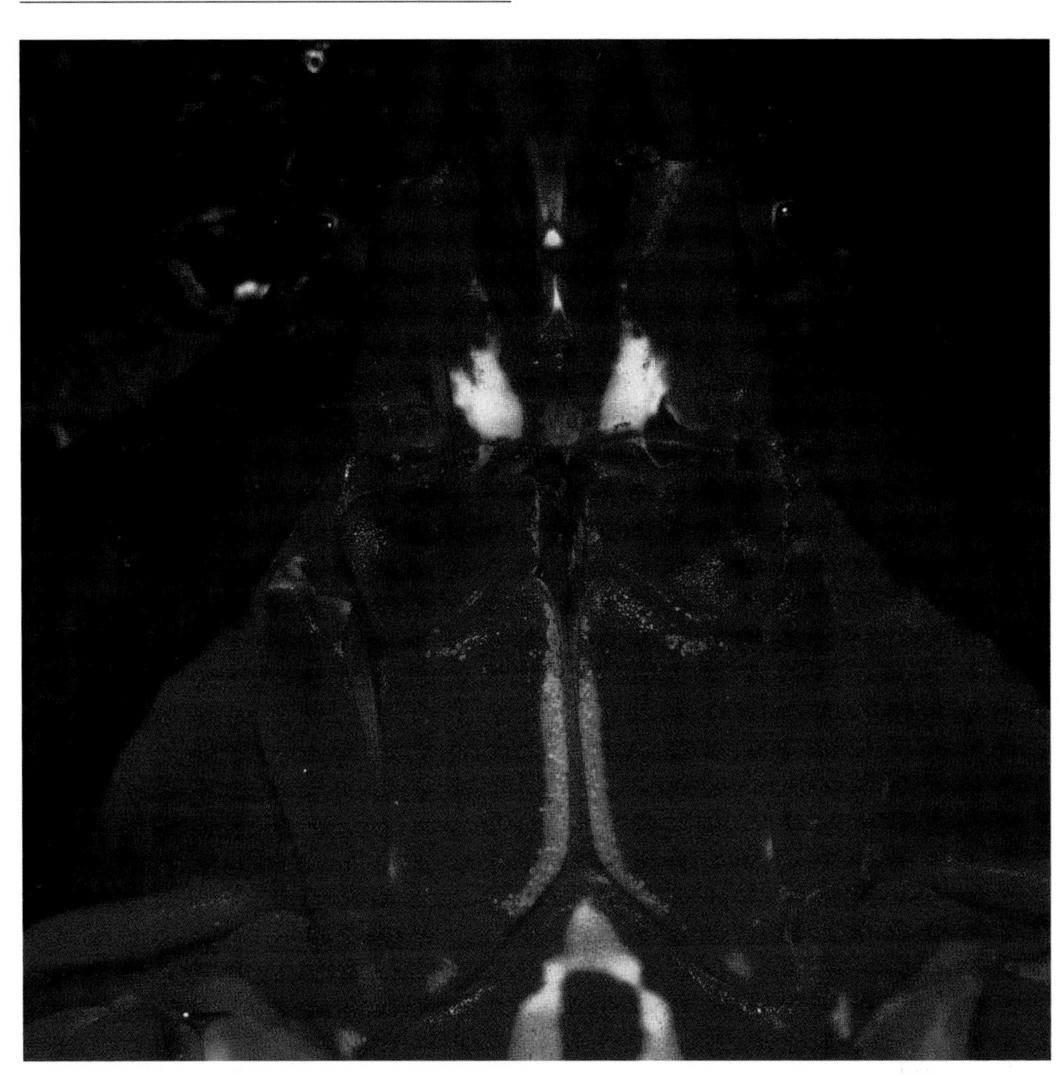

The underside of this northern kelp crab shows the detail of its armored mouthparts which, are used for chewing, macerating and tearing apart the variety of kelps and algae upon which it feeds.

terrace. It is narrow in comparison with worldwide averages and ranges from about 11 miles off Cape Blanco to 46 miles off the central coast. As one would expect, the shelf is steepest where it is most narrow. The depth of the shelf varies but is usually taken to be 600 feet, the average depth where the continental shelf turns downward sharply to become the continental slope.

Submarine Canyons

Submarine canyons represent some of the most dramatic changes in shelf topography. Off Oregon, the outer edge of the continental shelf and continental slope is dissected by two prominent submarine canyons and numerous smaller ones. The Astoria Canyon cuts into the outer shelf about 10 miles west of the Columbia River. During periods of lowered sea level, the Columbia and Rogue rivers drained across what is now the continental shelf. The Rogue Canyon is much smaller than the Astoria Canyon. It begins near the edge of the shelf offshore of the Rogue River and feeds directly down the continental slope onto the deep ocean floor. The Washington shelf is dissected by four major canyons: Juan de Fuca, Quinault, Grays and Willapa. These canyons and the troughs that feed them act as catch basins for sediments that originated on land and also for the rain of organic particles that is constantly falling on the sea floor. Most of this material is made up of the skeletons of plankton. Submarine canyons also play a significant role during upwelling events and nutrient flows.

These two juvenile box crabs are only about an inch wide but they will grow to over two feet when mature.

The Oregon Coast

The coast of Oregon can be characterized as having abundant rocky coastal areas, uplifted marine terraces and large beach features such as barrier islands. About 40 percent of the approximately 313-mile length of the Oregon coast consists of rocky sea cliffs and headlands, making it one of the most visually diverse coastal landscapes in the world. Much of the southern third of the coast consists of intensely folded and faulted sandstones, conglomerates and volcanic rocks dating as far back as the Jurassic (136 million years ago). At Coos Bay the coast abruptly changes, with the longest beach on the Oregon coast extending northward for 50 miles before reaching another headland, Heceta Head. Behind this beach is another accumulation of sand, the southern Oregon dunes.

With a face that looks like it got hit by a truck, wolfeels are territorial fish that mate for life. Male/female pairs generally live in one place for the greater part of their lives. This male is eight feet long.

The northern half of the Oregon coast consists of a series of beaches, separated by rocky headlands. Nearly all are composed of rock-hard basalt. Sand spits are common on the northern coast. Some point northward, others southward depending upon the action of winds, currents and tides. Many of the beaches are backed by sea cliffs cut into the uplifted marine terraces, exposing two million-year-old terrace sands as well as Tertiary marine mudstones and siltstones. Deposits from the shallow continental shelf, ancient beaches and dunes are represented in the terrace sands. Landslides are common in some parts of the central and northern Oregon coast as a result of these unstable sedimentary deposits and the high levels of precipitation for which the Oregon coast is known.

These stretches of beach are punctuated by bays and estuaries that offer little resistance to the significant wave action that occurs on this highly exposed coastline. In contrast, the rocky headlands seem to defy the constant pounding of the Oregon surf, but, of course they too will eventually fall victim to erosion. By interrupting longshore currents and wave patterns, headlands greatly affect the ability of sand to accumulate or be taken from the beaches between them. Because of this, these stretches of beach are considered pocket beaches, even though some of them may be rather large, extending for several miles. The primary source of sand for these beaches comes from sea cliff erosion of uplifted terraces. Larger coastal rivers such as the Siletz end in bays or extensive estuaries, which apparently prevent their loads of sediment from reaching the beaches on the outer coast.

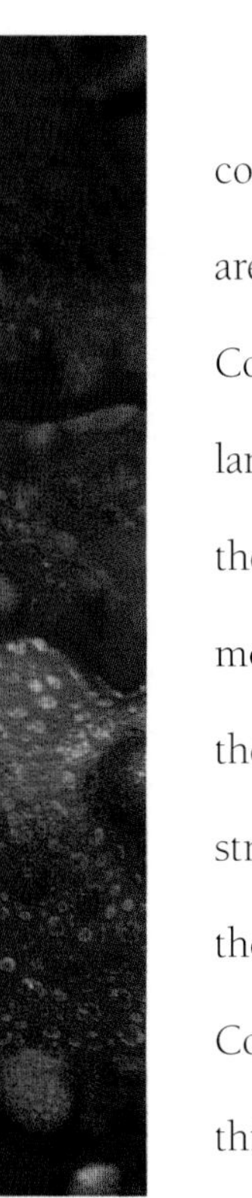

This profile of a mature box crab shows the well-armored claws and legs pulled up tight against the body to form an impenetrable box defense.

Beaches on one part of the Oregon coast are fed by a major river. This is the area extending southward from the Columbia River to the first major headland, Tillamook Head. Although most of the sand derived from the Columbia River moves northward into Washington during the spring/summer season, this 19-mile stretch of beach can actually be considered the southern portion of the "delta" of the Columbia River, so great is its influence on this part of the Pacific Northwest Coast.

The Washington Coast

The coastline of Washington is 2,664

miles long and consists of three distinctly different regions: the open, exposed coast, which meets the Pacific Ocean head-on; the northern shore of the Olympic Peninsula, including the Strait of Juan de Fuca and the San Juan Islands and the West Coast analog of Chesapeake Bay—Puget Sound. Each of these three areas of the state consists of different marine habitats housing different groups of marine life.

As in Oregon and northern California, the open coast of Washington is exposed to the constant crashing surf of the Pacific. One consequence of this ceaseless activity is the accumulation of heavy driftwood concentrations found along the shore, comprised mostly of western hemlock, Douglas fir and Sitka spruce, all of which make their way to the sea via the many rivers that empty into the Pacific. Washington tides have two unequal lows and two highs each day, ranging from 6 feet at neap to 12 feet at spring. Onshore winds are highly seasonal, varying between north-northwest in the spring and summer, and south-southwest in the fall and winter. While the California Current is most active offshore, inshore waters move slowly both south and north.

The Pacific Coast shore, between the Columbia River and the Strait of Juan de Fuca, can be divided into three geomorphic sectors: southern, central and northern. The southern sector is backed by a gently rolling plain with two large estuaries: Willapa Bay and Grays Harbor. Both were developed by the flooding of river mouths during a time of lower sea level. Northerly migration of the main

Alaska and British Columbia are home to a species of feather star. These delicate relatives of brittlestars normally sit and filter feed on the ocean floor. When disturbed they can "swim" away from an intruder.

This close-up image of the suckers of a giant Pacific octopus shows the detail of ribbing on the "skin" of each sucker, which is partially responsible for its ability to grasp so strongly.

Willapa Bay inlet channel is believed to be the cause of severe erosion at Cape Shoalwater, where the shoreline has receded 2 miles in the last 90 years. Two of the three major artificial structures along the state's outer coast are jetties that help stabilize the inlet entrance to Grays Harbor from shifting and eroding sands. The southern coast consists of broad, prograded sandy beaches fronting multiple rows of beach and dune ridges whose sediments originate from the Columbia River to the south. The construction of dams along the Columbia has impeded the flow of sediments to the sea, and over the last century it appears that sediments on the Columbia delta may be eroding rather than accumulating.

The central sector of the state consists of moderately wide sand and gravel beaches. These stand in front of cliffs made of continental and marine sedimentary rock produced during the Tertiary, between 5 to 60 million years ago. These strata are covered by glacial deposits delivered to the area over the past two million years as glaciers moved down from the north and receded back dozens of times. Much of this sedimentary material becomes eroded as a result of "mass wasting." This occurs when large chunks of sedimentary material enter the sea through landslides when cliffs become saturated with freshwater runoff. This material contributes significantly to the formation of central Washington coast beaches. For the most part, the central Washington coast forms a pretty straight line, which is connected to the southern sector by long stretches of exposed beaches. This part of the coast represents a transition zone between the south and north sectors.

The northern sector, including the Olympic Peninsula, is one of the most spectacular coastlines in the world. Here

Octopus are tremendous quick-change artists, and this species in Puget Sound, Washington, has become a mottled red color to blend in with the surrounding red algae.

the Pacific Ocean meets the West Coast in a display of huge, rolling ocean swells that crash into eroded rocky sea stacks, cliffs and headlands. Sediments do exist in a few places in the form of coarse gravel and cobbles. Sandy beaches are to be found only in the occasional pocket embayment along this stretch of the coast.

The Strait of Juan de Fuca is the major conduit for the flow of oceanic water into the inner reaches of the northern sector. It harbors a remarkably diverse collection of marine life. Bathed in clean, cold, Pacific Ocean water, yet sheltered from the constant pounding of ocean swells, the marine life along the strait has the best of all

Inhabiting subtidal sandy-mud bottoms, this juvenile hermit crab drags its hairy claws, legs and mouthparts through the sediment, sorting it for small bits of edible organic debris.

SOUTH SLOUGH

Habitats
Upland forest
Freshwater wetlands and ponds
Salt marshes
Tidal flats
Eelgrass meadows
Open water

Key Species
Port Orford cedar
Bald eagle
Great blue heron
Elk
Dungeness crab
Ghost shrimp

Description
South Slough features uplands of northwest coniferous forest and shrub, draining to freshwater and saltwater tidal wetlands, subtidal habitats and open water. Freshwater marsh areas resulting from historic agricultural dikes are to be restored to tidal influence.

Location
5 miles southwest of the city of Coos Bay, Oregon

worlds. This rich assemblage of marine life extends into the San Juan Islands (where it becomes somewhat less diverse) and up into the Canadian San Juans and the Inside Passage of British Columbia. Strong tidal currents are the main oceanographic influence in the strait and throughout the San Juan Islands, making navigation an exercise in caution and experience.

At the terminus of the strait is the entrance to Puget Sound, a deep-water estuary. Here the geology consists primarily of beaches and bluffs developed in Pleistocene times (the last 2 million years) mostly as the result of glacial activity. The sound acts a giant basin collecting freshwater runoff (and the sediments that come with it) from the enormous surrounding watershed that is the Puget Lowland. These sediments have created spits and bars of all sizes and deltas where streams and rivers enter Puget Sound. A major feature is Hood Canal, another deep-water estuary which

Red rock crabs go through a series of color "morphs" before achieving the adult red color. This all-white juvenile is trying to blend in with the white shell fragments of this sandy depression.

Gliding along on eleven uneven arms, a thorny sea star is probing the waters of southeast Alaska in search of anything slow enough to catch and eat.

This brown Irish lord is laying in a bed of pink soft corals, yellow sponge, and white metridium *anemones in the frigid waters of the Queen Charlotte islands of northern British Columbia.*

runs along the eastern side of the Olympic Peninsula for approximately 65 miles in length and 195 in depth. It is a glacial trough of enormous proportions that is now filled with water.

Alaska

Without a doubt, the coastline of Alaska is America's largest, most diverse and perhaps least known coastal region. It spans 20° of latitude and 50° of longitude. It has a general coastline of 6,625 miles and a tidal coastline of 33,830 miles, exceeding that of the entire continental United States. Four of North America's major physiographic divisions are part of

Voracious predators of other fishes, lingcod can grow to over five feet and 80 pounds in the Pacific Northwest. But, like the giant jewfish of the Florida Keys, most large lingcod have been fished out in heavily populated Puget Sound.

PADILLA BAY

Habitats
Open water
Seagrass meadow
Tidal flats and sloughs
Salt marsh
Upland forest
Upland meadow

Key Species
Seagrass (*Zostera marina* and *Zostera japonica)*
Dungeness crab
Salmon
Black brant
Bald eagle
Peregrine falcon

Description
Padilla Bay features extensive seagrass meadows and mudflats, channel subtidal habitats and fringing salt marshes. Agricultural sloughs flow into the bay and agricultural diking is a notable feature.

Location
5 miles north of Sate Highway 20 between Burlington and Anacortes, Washington

Alaska: the Interior Plains, the Rocky Mountain system, the Intermontane Plateaus and the Pacific Mountain system. All extend across Alaska's coast, and each has an important bearing on coastal geology and coastline character.

There are five geological and oceanic settings that make up Alaska's coast: the Arctic, Bering Sea, Aleutian Island, Gulf of Alaska and the southeast. The Arctic coast extends from Demarcation Point on the Canada-Alaska border west to Point Barrow and then southwest to Cape Prince of Wales. To the east of Point Barrow is the Beaufort Sea, to the west is the Chukchi Sea. From the Canadian border west to Point Barrow the coast is characterized by low cliffs and the coastal interruption of many small and large rivers that flow across the coastal plain. Some of these cliffs face the open sea, although most are separated from it by a series of shallow lagoons and low barrier islands. The western half of the Beaufort Sea coast is irregular in shape and is made up of a number of large bays. Ice-wedge polygons border much of the coast. Southwest of Point Barrow, tundra cliffs are present, varying in height between 30 to 54 feet. However, at Cape Lisburne they are more than 900 feet high. This is also the point where the Brooks Range meets the Alaska marine shore. Lengthy spits and barrier islands separated by narrow inlets occur along the middle section of the coast between Point Barrow and Cape Lisburne and along the northwest-facing portion of the Seward Peninsula. Nearly all of the Arctic coast is within the zone of continuous permafrost

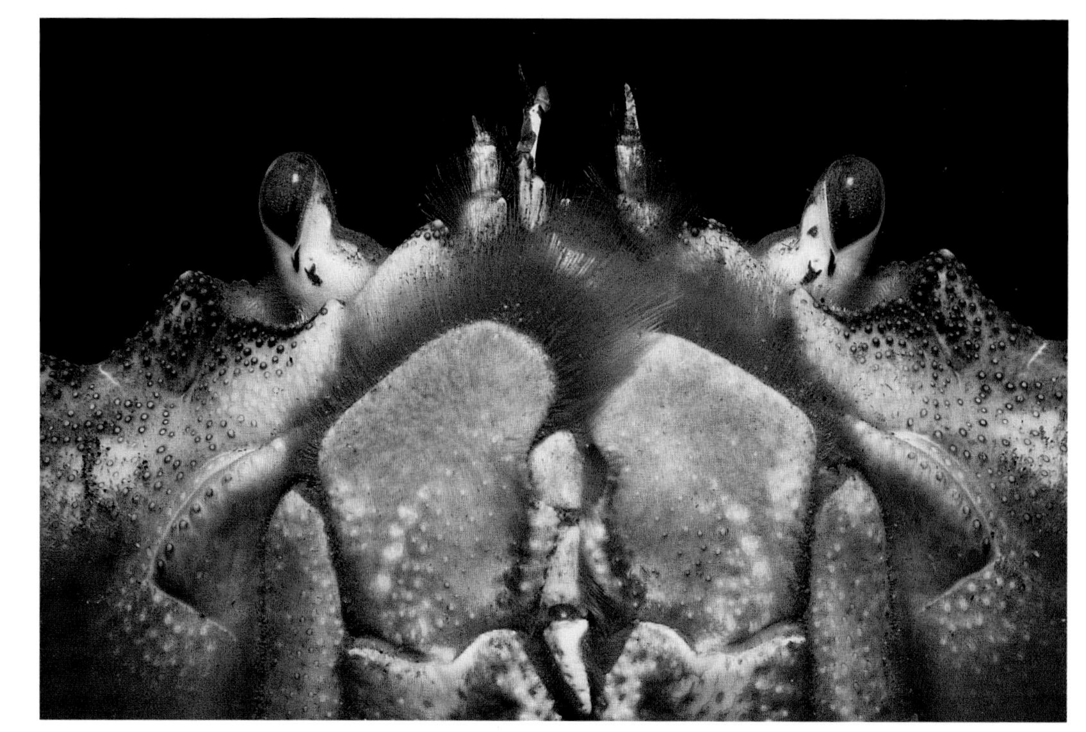

Eyes and mouthpart detail, Dungeness crab, Puget Sound, Washington.

and permafrost-related ice forms. Wedges, ice-wedge polygons and thaw lakes are common. Permafrost-related processes such as thermal melting and thermal erosion contribute to the overall appearance and deformation of the coast in this area. Differential thaw and erosion often produce undercuts and serrated, low-cliff coastal features. When tapped by ocean waves, thaw lakes become part of the shallow foreshore, creating a coastline that is arcuate in form.

Norton Sound to the north and Bristol Bay to the south are the dominant coastal features on the the Bering Sea coast between Cape Prince of Wales and the western end of the Alaska peninsula. These enormous embayments are separated by the extensive Yukon-Kuskokwim deltaic complex and the southwest extension of the Kuskokwim Mountain Range. Norton Sound is characterized as having a relatively smooth coastline with narrow beaches backed by glaciated hills. The Yukon-Kuskokwim delta consists of many lakes, marshes and abandoned tributaries. Presently the Yukon River, which contributes the majority of the sediment that reaches the Bering Sea, discharges north into Norton Sound, where intertributary

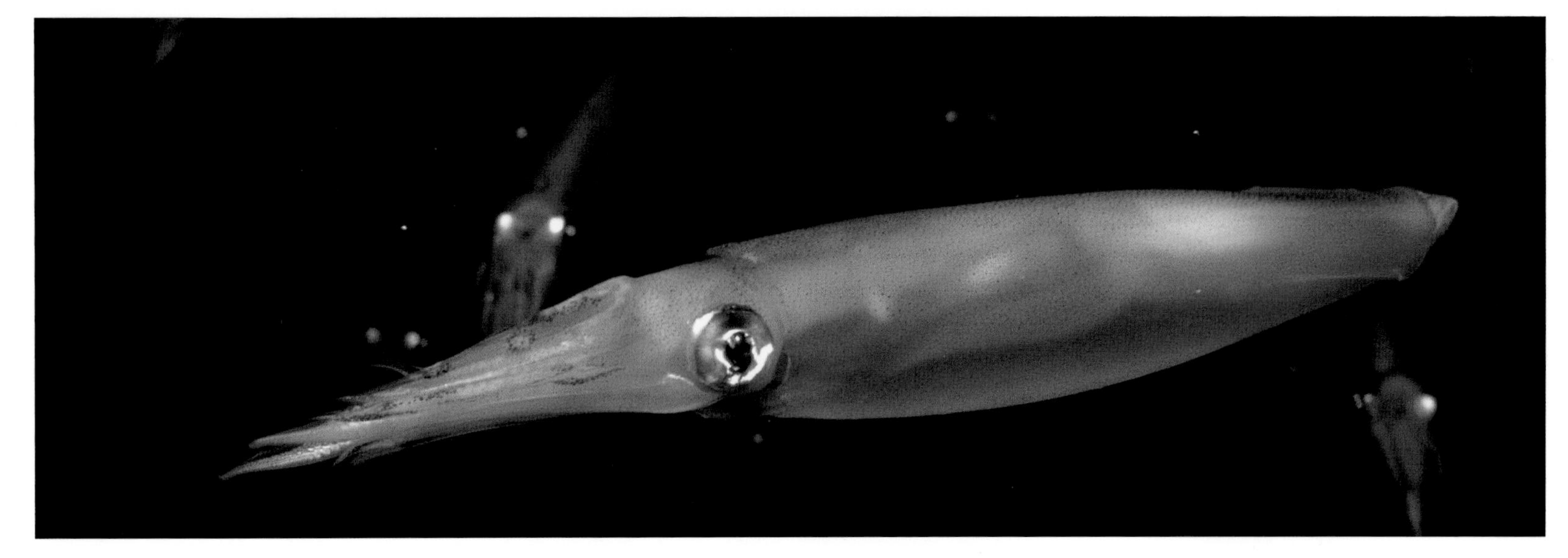

The common squid, Loligo opalescens, *is a member of the open ocean crowd along the entire West Coast. Unlike their relatives the octopus, these cephalopods are streamlined and finned for open water cruising.*

A common member of Pacific Northwest plankton, this small jelly can be found in abundance in the spring and summer. The small, red spots at the base of its bell are eye spots that register only light and dark.

mudflats are extensive. These become homes for a variety of infaunal organisms such as polychaete worms, clams and burrowing amphipods.

In the southwestern part of the delta complex, gravel beaches and rocky headlands are numerous where volcanic activity formed hills composed of this rock. Nelson Island, near the delta and Nunivak Island, about 31 miles offshore, have similar coastlines. In contrast, much of the coast of Saint Lawrence Island has low sandy beaches, some of which are even backed by sand dunes. South of the Kuskokwim Estuary the Kuskokwim Mountains meet the sea with high, rocky cliffs and numerous sea stacks, creating an impressive landscape. Most of the Bristol Bay coast is composed of glacial and fluvial sediment that have created a variety of soft-surface habitats such as sandy beaches, extensive mudflats and numerous lagoons.

The Aleutian Islands extend in a gentle arc for nearly 1,240 miles across the northern Pacific Ocean to Attu Island at its western extremity. These islands are the exposed portions of a 10-to 62-mile-wide ridge that flanks the western half of the Aleutian Trench, which is over 1,860 miles long and up to 22,500 feet deep. This arc represents the subducting northern edge of the Pacific Plate where it meets the northern portion of the American Plate. Along with this active subduction are frequent earthquakes in southern Alaska. There are 60 centers of volcanic activity in the region that extend from near Anchorage to the western end of the Aleutian Islands. Of these 60 centers, more than 40 have been active since 1760. Many of the coasts in the Aleutian Islands are composed of lava that flowed into the sea or of ash that has fallen on the shore or into the water and then washed on shore; other coasts have been eliminated or altered by explosive volcanic eruptions. Throughout the islands the coasts have rocky cliffs that alternate with boulder beaches. In some protected coves sandy beaches are present. Abrasion platforms are found at present-day sea level as well as at various heights below and above sea level (up to 600 feet) as a result of tectonic uplift.

The coast of the Gulf of Alaska, which extends in a clockwise direction from the eastern Aleutian Islands to Cape Spencer, includes a number of distinctive coastal types. The Pacific-facing portion of the Alaska Peninsula is highly irregular and has numerous rocky headlands that alternate with formerly glaciated inlets. Many of the headlands are continued offshore by rugged islands and rocky stacks. One of the largest islands in Alaska is Kodiak, which is separated from the northern part of the Alaska Peninsula by a 31-mile-wide strait, and it too has an irregular and rugged coastline with fjords, especially along its northwestern coast. Cook Inlet, a deep structural basin over 186 miles long, is a tidal estuary that opens into the Gulf of Alaska and has some of the world's highest tidal ranges (30 to 33 feet). Near its head extensive tidal flats and marshlands occupy the coastal edge of a low coastal plain, which narrows gradually on the northern side of Cook Inlet and toward its mouth merges into mountains that descend directly into the sea. In contrast, on the southern side of Cook Inlet pocket beaches alternate with low cliffs along most of its extent.

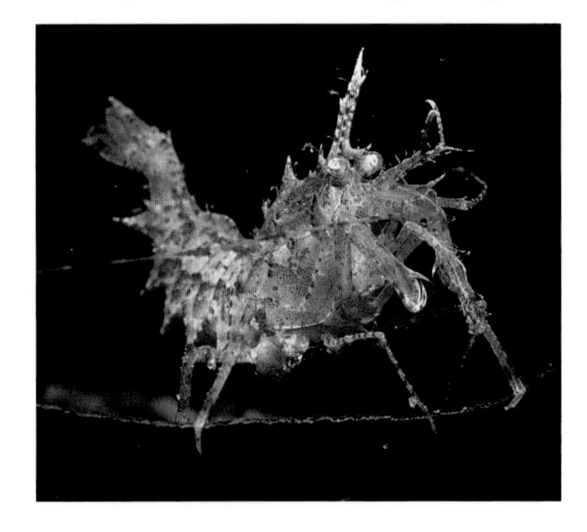

Looking like something out of a sci-fi horror film, this Paracrangon *shrimp is an active predator of other, smaller crustaceans, worms and small snails in the cold waters of the Pacific Northwest.*

The coast from the mouth of Cook Inlet east around Prince William Sound is deeply indented with fjords. Much sediment is transported into these fjords by the short, turbulent streams that originate in the alpine glaciers that are still present in the coastal mountains. From Prince William Sound southeast to Cape Spencer, the coastline is smooth in outline and is characterized by deltas, beach and dune ridges, outwash plains, moraines and glaciers. In some locations tidewater glaciers are present, which represent some of the most rapidly changing coastlines to be found in Alaska.

The southeast Alaska coast extends from Cape Spencer south to Tongass at

Dixon Entrance on the Alaska-Canada border. Although the general length of the southeast Alaska coast is only 285 miles (4.3 percent of Alaska's total), its tidal shoreline is 12,729 miles long, or 37.6 percent of the total for the state. Such a variation in its general and tidal coastline lengths is due to the large number of inlets, straits, bays, fjords and islands that characterize this rugged region. Many steep cliffs of these fjords descend hundreds of feet into the ocean. However, there are low coastal plains near the head of some of the fjords. Strand flats and reefs border many islands and headlands, especially those exposed directly to the ocean.

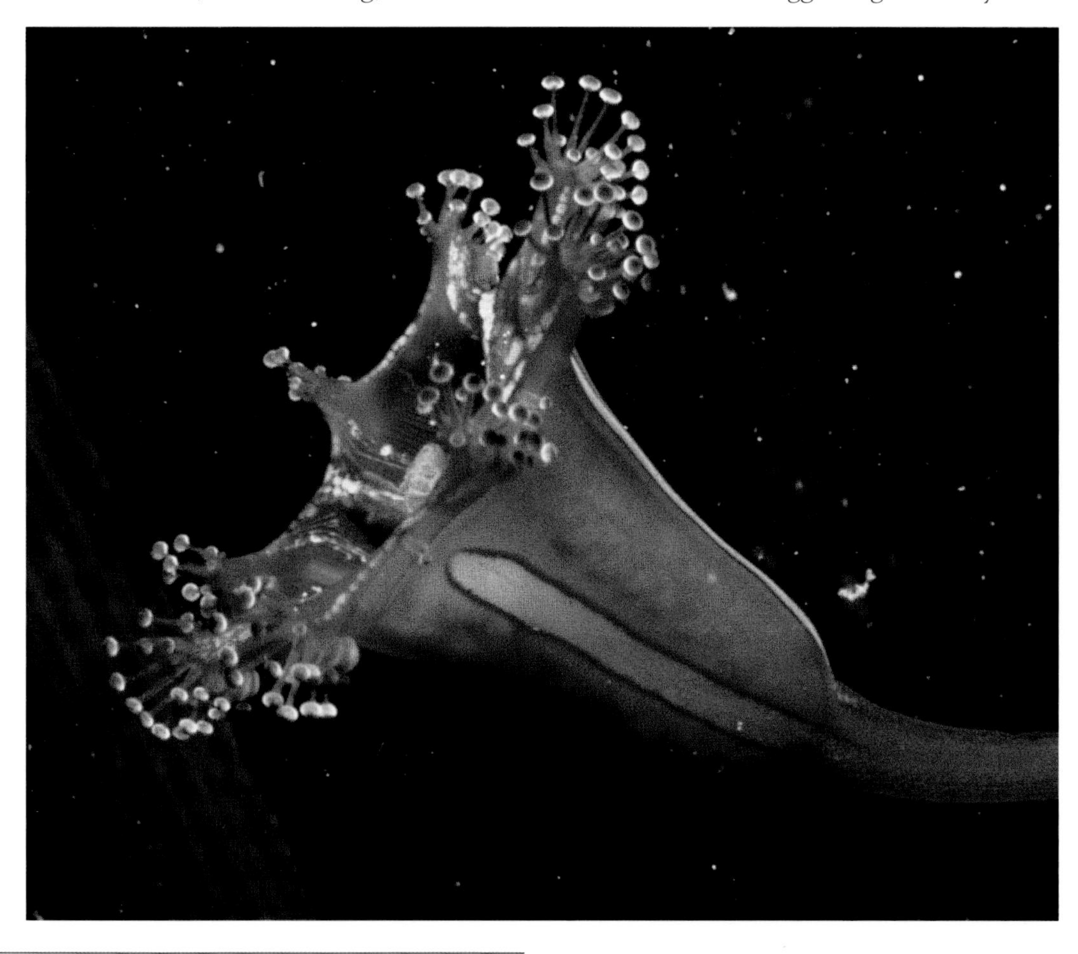

Doing a sea anemone imitation, this "jellyfish" has adopted a sedentary mode of existence, attaching itself to eelgrass and other algae with a singular suction disc.

Most of the coastline is protected from the open sea, but tides of at least 7 to 21 feet in some of the fjords and bays are responsible for strong tidal currents and intensive scouring action. In contrast, the gross geomorphic character of southeast Alaska is the result of the glacial modification of a tectonically active, complex mountainous terrain.

Pacific Northwest Zoogeography

The Pacific Northwest and southeast Alaska can be divided into two main provinces; a southern "Oregon" province and a northern "Aleutian" province. The Oregon province extends north from Point Conception to the Alaska Peninsula, encompassing a very rich, cold, temperate fauna. The exact location of the northern boundary has not been clearly defined, and varies with certain groups of marine life. The most diverse starfish fauna in the world extends north to about Sitka. The algal flora is very uniform from the mouth of the Columbia River as far north as Sitka. The rich shorefish fauna found north of Point Conception extends into the cooler waters of British Columbia but rapidly changes character as it reaches southeastern Alaska. In northern California roughly half of the shorefishes are endemic. The rockfishes alone in the genus *Sebastes* form a large complex of more than 50 related species. More than two-thirds of these species do not extend north of British Columbia or southeastern Alaska. Of the *Sebastes* species that occur in British Columbia, the majority do not extend as far north as the Gulf of Alaska. In fact, about 50 percent of the entire shorefish fauna of western Canada does not extend north of the Alaskan Panhandle. The demersal fishes of economic importance exhibit the greatest change in species groups within the British Columbia, southeastern Alaska, and Gulf of Alaska area. Many Bering Sea species also extend into the Gulf of Alaska but not further south. These range distributions all indicate the presence of an important zoogeographic boundary defining the northern-most distribution of many species along southeast Alaska. When the fishes, echinoderms and marine algae are all examined, the northern boundary of the Oregon province seems to be somewhere along the Alaskan Panhandle, perhaps as far north as Sitka.

Nudibranchs, better known as sea slugs, are among the most colorful and engaging animals of the Pacific Northwest and Alaska. The long, finger-like projections covering this species' back are its gills.

The Aleutian province lies between the Oregon province and the really cold waters of the Arctic region. The northernmost limit of this province lies somewhere

OLYMPIC COAST

Habitats
Rocky and sandy shores
Kelp forests
Seastacks and islands
Pelagic, open ocean

Key Species
Tufted puffin
Bald eagle
Northern sea otter
Gray whale
Humpback whale
Pacific salmon
Dolphin

Description
Spans 3,310 square miles of marine waters off the rugged Olympic Peninsula coastline. The sanctuary averages approximately 35 miles seaward, covering much of the continental shelf and protecting habitat for one of the most diverse marine mammal faunas in North America and a critical link in the Pacific flyway. The sanctuary boasts a rich mix of cultures, preserved in contemporary lives of members of the Quinault, Hoh, Quileute and Makah tribes.

Cultural Resources
Native American petroglyphs and villages
Historic lighthouses
Shipwrecks

Location
From Cape Flattery to the mouth of the Copalis River, on Washington's outer coast

Protected Area
3,310 square miles

near the Bering Strait and, as with the Oregon province, identifying a well-defined boundary for all marine flora and fauna is difficult. This is primarily because of the different ways various marine organisms disperse their young, and the degree of mobility achieved by adults of different animal groups. Obviously fish are more mobile than snails, and indeed it would appear that a fair number of Bering Sea fishes do periodically invade the north-central coastal regions of Alaska. In any event, once again water temperature seems to be the overriding physical factor determining the distribution of marine life in this area. Studies of distribution and occurrence of mollusks suggests the northern limit may be near Nunivak Island. This also corresponds with the southern limit of pack ice in January and February, representing a clearly defined boundary in terms of water temperature.

This sea slug is known for its ability to eat sea anemones, including their stinging cells. As the final insult, it lays its eggs on whatever is left of its prey.

Giant sponges, sea anemones and even soft corals reminiscent of the tropical Pacific adorn a sheer drop-off in the Queen Charlotte Islands south of Alaska.

Little is known about the details of the distribution of fishes along the eastern coast of the Bering Sea. It appears that many Arctic species enter the northern Bering Sea but these are not found in its southern reaches. Conversely, a large number of north boreal Pacific species range up into the southern Bering Sea but are rare or absent in the northern end. Most of the bottom fishes reported from Bristol Bay appear to be species that do not occur in the northern Bering Sea. Since there is so little information available about distributions within the Bering Sea, it would be risky to hazard a guess as to the general degree of endemism that exists in the Aleutian province. It does appear that the fauna is less rich than that of the Oregon province immediately to the south, and since the geographic area is also considerably less the endemism is probably less as well.

Relationships with the North Pacific Rim

The Bering Sea represents the zoogeographic link to the northwestern portion of the Pacific Rim, and the gateway to Asian marine life provinces. The Bering Sea is essentially a broad, shallow basin almost completely enclosed to the north while its southern end is bordered by the Alaska Peninsula and the Aleutian chain of islands. In considering this general topography, the absence of obvious barriers might lead one to expect a homogeneous marine fauna. However, recent investigations have shown that the eastern and western parts of the southern Bering Sea (below the Arctic zone) have quite different faunas.

Even though the Arctic waters apparently come far enough south to form a temperature barrier at about Nunivak Island on the east and Cape Olyutorsky on the west, it is still surprising to find a high degree of difference between the two sides of the southern Bering Sea, given that this barrier is equal for both sides at this latitude. While the distance between Nunivak Island and Cape Olyutorsky is only about 800 miles, about half this distance is covered by the shallow waters of the continental shelf. It may be that for many of the shore species of invertebrates and fish the deep waters of the northern Aleutian Basin off the Siberian coast form a significant barrier to any westward migrations.

Further south the relationship between the eastern and western sides of the north Pacific Rim is much less marked. Looking at a map it might appear that the Aleutian chain would provide an easy migration route for dispersing fauna between Russia, Japan and the American west, much the same way the islands of the South Pacific act as stepping stones for the transport of marine life to Hawaii. But this does not seem to be the case given the low numbers of related species from both sides. For example, only about 0.3 percent of mollusks have used this route. Other organisms, such as hermit crabs and some starfish, have made it across via the Aleutian chain. There are also several arctic-boreal species that have continuous or nearly continuous ranges through the northern Bering Sea extending far to the south on each side of the North Pacific. The presence of such species accounts for most of the similarity that can be found between the marine fauna of the eastern Oregon province and what is known as the Kurile province on the western side of the north Pacific. Examples among the fishes include the starry flounder (*Platichthys stellatus*) and five Pacific

One of the Pacific Northwest's more flamboyant fishes, the sailfin sculpin sculls the water with its enlarged dorsal fin, using the stiff forward portion as a rudder.

salmon species belonging to the genus *Oncorhynchus*. For the echinoderms there is the giant red sea urchin (*Strongylocentrotus franciscanus*) and a brittlestar (*Amphipholis pugetana*). Some of the shorebirds such as the tufted puffin (*Lunda cirrhata*) exhibit very similar distributional patterns. The Steller sea lion (*Eumetopias jubatus*) is a good example of a marine mammal that is found in both the Oregon and Kurile provinces.

The differences in the marine life of the two coasts expresses itself primarily at the species level. If we look one notch lower on the similarity scale—the genus level—there is significantly greater similarity between the faunas of these two shores. This level of similarity indicates the presence of two evolutionary centers of marine life, one on each side of the boreal Pacific.

With sticky, tendril-like arms unfolded, this basket star filter-feeds for plankton in the waters of southeast Alaska.

For example, the clam genus *Mya* and some other molluscan groups apparently underwent considerable evolution in the Kurile province and Okhotsk Sea areas. This is probably also true for most of the sea squirt (*Ascidian*) genera and certain fish families such as the sculpins (*Cottidae*), snailfish (*Liparidae*) and the eelpouts (*Zoarcidae*). In these groups, the majority of the species and most of the endemics are concentrated on the Asiatic side. The American side appears to be the evolutionary center of radiation for invertebrates such as certain crabs and shrimp (*Decapoda*), the asteroid starfishes and fish families including smelt (*Osmeridae*) and rockfish (*Scorpaenidae*).

Space is at a premium on rocky reefs in British Columbia and southeast Alaska, and every square inch is used by encrusting organisms. Here vermillion soft corals surround a crimson anemone, flanked by sponges, barnacles and sea squirts.

Underwater Wilderness

SEVEN

Hawaii & American Samoa

This fish-eye view of a crashing wave on the south coast of Lanai provides a different perspective of Hawaii's rugged volcanic shoreline.

Facing page: In the winter humpback whales "hang out" in Hawaii to breed. In the spring they will migrate back to Alaska, to feed.

The Hawaiian Islands make up a long archipelago that extends approximately 1,500 miles across the central Pacific Ocean from the Big Island of Hawaii in the southeast to tiny Kure Atoll at the northwest. Beyond the Kure Atoll chain lie the Emperor Seamounts. This series of underwater mountains extends all the way to the Aleutian Trench (approximately 2,200 miles from Hawaii) and is geologically related to the emergent Hawaiian Island chain. Six main islands and two lesser ones make up 99 percent of the land area of the 50th state. These are Hawaii, Maui, Kahoolawe, Lanai, Molokai, Oahu, Kauai and Niihau. The remaining 1 percent is represented by small islets, rocks and scattered patches of coral reefs appearing as atolls found in the northwestern

Hawaiian Islands from Nihoa to Kure.

These islands and seamounts represent the emergent crests of huge underwater mountains that originate roughly two miles deep from the floor of the Pacific basin. In fact, the island of Hawaii, if measured from the sea floor, is the tallest "mountain" on the planet, rising vertically more than 29,000 feet. Thirteen thousand six hundred and seventy-seven feet of it breaks the surface of the Pacific. Comprised of approximately 10,000 cubic miles of volcanic rock, it is the largest single volcanic structure on Earth, 100 times larger than Mt. Shasta or Fujiyama. On a galactic level, it is the biggest such feature in the solar system between the sun and Mars. And it is still growing. The current volcanic activity on Hawaii provides us with present-day evidence of the volcanic origin of these remote islands.

An aerial photograph of the islands clearly illustrates their mountainous character, and their differing rates of erosion, the primary evidence that they have emerged from the sea sequentially. Fifty percent of the state land area lies above an elevation of 2,000 feet, and 10 percent lies above 7,000 feet. The maximum elevations of the six major islands are: Hawaii, 13,796 feet; Maui, 9,200 feet; Kauai, 4,700 feet; Molokai, 4,550 feet; Oahu, 3,700 feet and Lanai, 3,600 feet. Approximately 1,000 miles of collective shoreline rims the major islands, only about 20 percent of which is composed of sandy beaches. The remainder of the shore consists of outcrops or boulders of lava punctuated with an occasional muddy shore, gravel beach, raised beachrock reef, or calcareous bench. Hawaiian shores vary in elevation from raised bench reefs only 3 to 6 feet above sea level to the spectacular sea cliffs along the Napali coast of Kauai, which rise from sea level to 1,800 feet high.

A sandbar shark swims effortlessly in the clear blue waters of the Auau channel between Lanai and Maui.

The Hawaiian islands originated as volcanoes. This chain of mid-ocean volcanoes has, over a period of millions of years, been built up from the sea floor by innumerable eruptions around a central hot spot, creating a sequence of island building from the northwest to southeast. The oldest volcanoes are the submerged mountain peaks forming the base of the

"Leeward Islands" in the northwestern portion of the archipelago. The youngest volcanoes are those on the island of Hawaii, which are still erupting. Ten to 20 million years ago the Leeward Islands were probably high islands that physically resembled today's main islands. Over time, these ancient volcanoes have succumbed to the power of erosion and subsidence, sinking into the sea and becoming overgrown with coral which could keep up with the rate of sinking. In geologic time, the creation of the "Aloha State" has been very recent, keeping in mind that the age of the Earth is roughly 4.5 billion years.

Schooling fish on a shallow reef off Molokini Crater, Hawaii.

Located in the middle of the vast Pacific Ocean, Hawaii is 2,000 miles from California and 3,400 miles from Japan. The islands' nearest neighbors are Johnston Island, 450 miles to the southwest, the Line Islands 1,000 miles south, and Wake Island 1,200 miles to the east. This physical isolation from the continental land masses of the Pacific Rim has been the primary factor in determining the way all organisms—land and sea—have evolved in Hawaii. It is the reason for two significant aspects of Hawaiian flora and fauna. The first is a high degree of endemism—the development of species found nowhere else in the world. The few species that "made it" to Hawaii, either by air or water currents, were the basic genetic stock from which many other Hawaiian species evolved. Until humans set foot on Hawaii, these species remained relatively isolated, and the source of new species was primarily through natural selection and the diversification of original stocks. This has made much of the flora and fauna of Hawaii unique. The second important aspect related to isolation is the high degree of ecological fragility in the islands resulting from the long evolution of a few species in a setting not subjected to competition by new species arriving from the outside. Humans have continually brought in new species at a rate that far exceeded the ability of the ecosystem to adapt. This explains, in part, why the introduction of new species to Hawaii has destroyed much of the unique land plants and animals as well as caused serious problems for their aquatic counterparts. Humans have not been exempt from susceptibility to outside organisms, as evidenced by the death of many native Hawaiians when they were exposed to diseases such as syphilis.

Getting to Hawaii has not been easy for any organism because of its isolation. Weather and oceanographic patterns are the primary transport agents to the islands, either by wind or water current patterns. Further, both its location and regional climate have had and continue to have a profound effect on the kinds of organisms that can live in Hawaii. Hawaii is the only state that lies within the tropics, although just barely so. In fact, it is really considered a subtropical zone, lying just south of the Tropic of Cancer. It is also the only state in the union made up entirely of relatively small islands that are completely surrounded by salt water. This convergence of water, wind and land cre-

Like many butterflyfish, these raccoon butterflies travel in mated pairs. The black band running through the eye helps confuse predators.

ates Hawaii's weather, and in turn affects its plant and animal life.

During most of the year the northeast trade winds account for the dominant air movements over the state, and rainfall distribution is influenced primarily by the interplay of the trade winds with the islands' terrain. As warm, moist air from the Pacific meets the islands' peaks, it dumps massive quantities of fresh water in the form of rain on these windward areas. From May through September, the trades are prevalent 80 to 95 percent of the time. From October through April, the trades are prevalent only 50 to 80 percent of the time. On the average, major storms occur from two to seven times per year, usually between October and March. It is during these winter storms that the dry leeward lowlands receive most of their annual rainfall. Hawaii is also subject to occasional hurricanes, most recently Hurricane Iniki, which demolished much of Kauai.

Because of its location, Hawaii experiences very slight seasonal changes in its weather. The nearly constant flow of the trade winds bringing fresh ocean air of relatively uniform temperature into the islands also contributes to minimal changes in air temperature in the islands. Day length in Hawaii is relatively uniform throughout the year and the small annual variation in the altitude of the sun above the horizon results in small variations in the amount of incoming solar energy. While these climatic conditions may be the primary reason vacationers flock to the beaches of Waikiki, they have also had a profound effect on Hawaii's flora and fauna over time. Unlike environments in higher and lower latitudes, Hawaii has enjoyed this moderating climate for millions of years, probably since its beginning. Thus the organisms living there have historically not had to deal with a harsh or changing climate (although Hawaii's large volcanoes do have more pronounced "seasons"). In biogeographic terms, stable climates encourage the diversification of species, as biological factors such as competition become a primary force in the process of natural selection.

While the average annual rainfall of the state is about 70 inches, it falls at extremely different rates at specific locations depending upon the nature and orientation of the terrain. At Kawaihae on the leeward coast of the Big Island, the average annual rainfall is less than 7 inches; at Mount Waialeale on Kauai it is 500 inches.

Triggerfish are easily identified by their nearly symmetrical body shape and their use of dorsal and anal fins for propulsion.

HAWAIIAN ISLANDS HUMPBACK WHALE

Habitats
Humpback whale breeding, calving and nursing grounds
Coral reefs
Sandy beaches

Key Species
Humpback whale
Pilot whale
Hawaiian monk seal
Spinner dolphin
Green sea turtle
Trigger fish
Cauliflower coral
Limu

Description
The shallow, warm waters surrounding the main Hawaiian Islands constitute one of the world's most important humpback whale habitats. Scientists estimate that two-thirds of the entire North Pacific humpback whale population migrate to Hawaiian waters each winter to engage in breeding, calving and nursing activities. The continued protection of humpback whales and their habitat is crucial to the long-term recovery of this endangered species.

Cultural Resources
Native Hawaiian traditional practices
Native Hawaiian fish pond
Archaeological sites
Historic shipwrecks

Location
Within the 100-fathom isobath in the four island area of Maui, Penguin Bank and off Kilauea Point, Kauai.

Flash floods are not uncommon in Hawaii during intense, but brief, periods of rainfall. The impact on marine organisms living along the shore can be great when large amounts of fresh water and sediments are dumped quickly into shallow bays and inlets. While this may represent a brief (but significant) change in temperature and salinity, rapid flushing of oceanic water brings these coastal environments back to their normal state fairly quickly.

Strictly nocturnal, glasseyes can be found during the day hiding in the deep recesses of caves or in one of the many submerged lava tubes found in Hawaii.

Situated in the zone of the northeast trade winds, surface water currents in the vicinity of Hawaii are the large, westward-flowing Eastern Pacific Gyre and the smaller, eastward-flowing Equatorial Countercurrent. For the most part these currents do not favor the migration and establishment of marine life from the species-rich Indo-West Pacific into the central and eastern Pacific. This is demonstrated by the number of marine species you can find across the Pacific, decreasing as you move from west to east. As always, there are exceptions, and a few species have successfully made the trip to Hawaii, and even all the way to the west coast of North America. A few species indicate relationships with Japanese species suggesting that the eastward-moving Kuroshio and North Pacific currents have delivered larvae to Hawaii from the southwest via Johnston Atoll and the Line Islands. The trip of these pioneer larvae may have been facilitated by the presence of submerged seamounts between the Marshalls and Hawaii that were at one time above or close to sea level, making them available as stops along the way.

This female spotted trunkfish was photographed in two feet of water on a shallow night dive on Maui. Males flaunt gaudy blue and yellow markings intermingled with their coat of spots.

Although the average water temperature during the summer can be as high as 80°F, winter surface temperatures can drop to 68 to 75°F. For a large part of the year this puts the waters of Hawaii near the minimum water temperature necessary for coral growth. Add to this the isolation factor mentioned earlier and you can see why the coral diversity of Hawaii is relatively low, especially when compared to the rest of the tropical Pacific. These relatively cold winter temperatures also have limited the variety and abundance of sponges, bryozoans, encrusting worms and other invertebrates found in Hawaiian waters.

In a characteristic pose, this blackside hawkfish surveys the reef scene off southern Maui. All hawkfish are sitters, using their pectoral fins as elbows.

Coastal Conditions

The Hawaiian Islands are the exposed tops of large volcanic mountains built up from the floor of the Pacific Basin. These volcanoes formed over a hot, localized spot in the earth's mantle under or just to the south of the island of Hawaii where magma rises and breaks through the Earth's crust. As these great shield volcanoes erupted, built up and died, the motion of the ocean floor and crust relative to the stationary hot spot moved them to the northwest.

One of the most interesting aspects of Hawaii's natural history is that you can see how the islands' shores have evolved. There is no mistaking the contrast between the cooling lava shores of Hawaii and the dense floral populations of the eroded coast of Kauai. At one end of this island chain there is a new coast being built while at the other end coral reefs are replacing lava as the islands sink slowly in the northwest.

Once lava enters the sea it immediately becomes subject to the processes of erosion and deposition, the two forces ultimately responsible for the shaping of all coastlines. The action of wind and water will break down the volcanic material into smaller and smaller particles, most of which will eventually be swept into the deep ocean. Deposition will lead to the accumulation of these eroded particles in places along the shore that are not subject to wave action and currents that will remove them from the shore. In Hawaii, the most distinctive beaches are those that are composed of the pulverized volcanic fragments that are deposited in coves or

Dragon morays are Hawaii's most distinctive moray eels. The twin projections on the tip of its upper jaw are extensions of its nose, improving its ability to sniff out prey.

One of Hawaii's larger morays, this yellowmargin is resting in a ledge that is obviously too small to accommodate its entire body.

How did the common name of Plectorhinus chaetodonoides *become "harlequin sweetlips"?*

other sheltered locations and form the unique black sand beaches especially evident on the big island of Hawaii.

As is to be expected, the newly formed shoreline of Hawaii displays less diversity and "character" than the much older coastlines of Kauai or Oahu. With time, additional geological processes have created a greater variety of structures, and biological agents such as corals have made their contributions to the shore as well. These include the development of reefs, additions of land sediments brought to the coast by streams and changes in sea level. Over a million years or so of gradual change, the blank, low cliffs along a coastal lava flow are replaced by eroded rock outcrops, wave-cut benches, high cliffs and sand beaches.

Beaches and Lava Flows

As mentioned earlier, only about 20 percent of Hawaii's shoreline is covered by beach sediments. This is primarily due to the lack of large rivers supplying sediment and the young age (geologically speaking) of the islands. The transport of sediments to deep water offshore is also facilitated by exposure to the tremendous ocean waves that continually pound Hawaii's shores, a situation that is exacerbated in some locations by local topography and bathymetry. The older islands have higher percentages

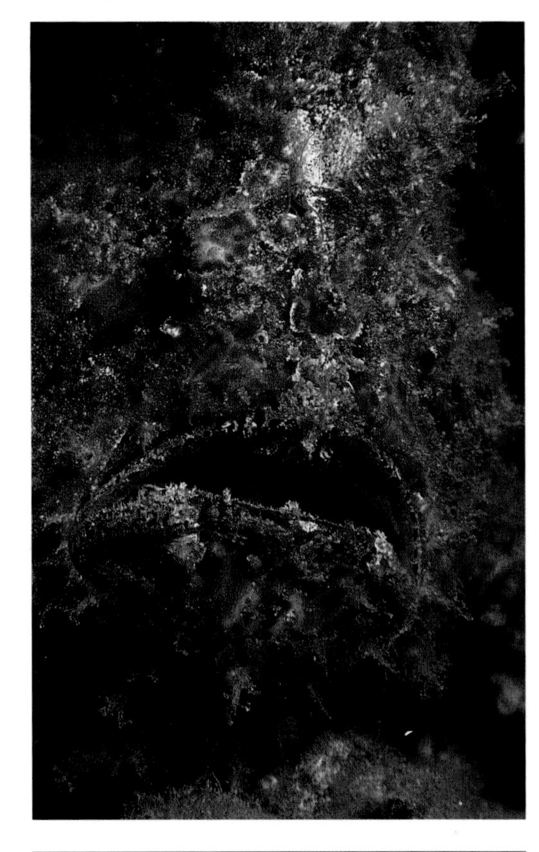

Only a slightly gaping maw and a twinkling eye give away the presence of a giant anglerfish on a shallow reef in Hawaii.

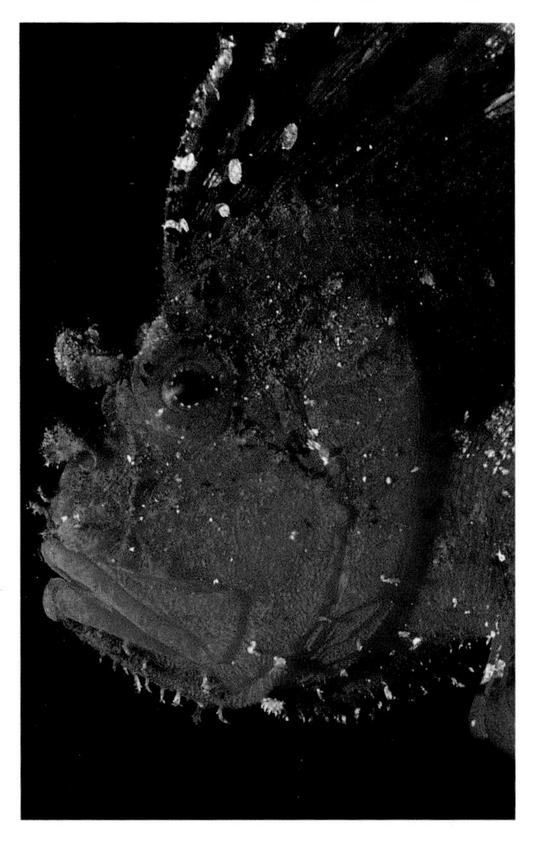

Leaf scorpionfish use camouflage for both defense and as a feeding strategy. Small and thin, they sit on the bottom and sway back and forth with the surge of waves, imitating a piece of drift algae.

of sandy shore than the younger ones.

The sand on Hawaiian beaches also differs from most continental beach sand in being principally calcareous and of biological origin. Most of the beach sand originates as shells of animals that live on the fringing reefs or in shallow waters near the islands. There are two exceptions, however. First are those beaches near stream mouths, made up of detrital basalt sand. Second are the beaches of black volcanic glass sand. These resulted from steam explosions at the chilled glassy margins of hot lava flows that extend to the ocean and were broken down by wave action.

Marine Biogeography

Hawaii's marine life is one of the most well-studied groups anywhere. There are several reasons for this. First, because of its isolation it provides ecologists and biogeographers with an opportunity to look at how island communities develop and evolve. Second, information on the

Slate-pencil urchins are one of the most common relatives of seastars in Hawaii. This view provides a close-up of the underside, revealing a mouth, tube feet and detail of the spines.

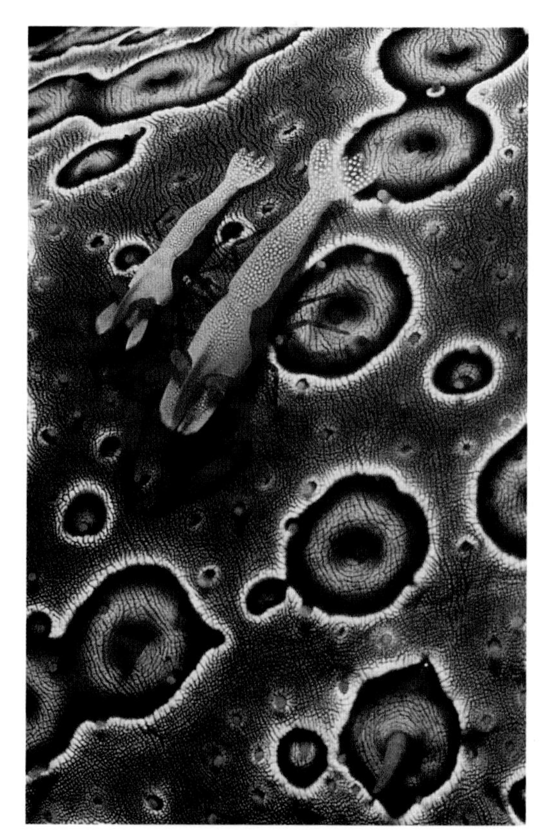

Appearing more like alien beings than Indo-Pacific marine life, this pair of symbiotic shrimp make their full-time home on a sea cucumber.

ecology of the islands can be compared with its geology and geography, both of which are very well documented. Bringing these two areas of study together has given us a more complete picture of how these kinds of island ecosystems work.

Where did all this marine life come from? For the answer, look to the west because the greatest concentration of marine life anywhere on the planet is found in the waters of the Indonesia-Malay Archipelago of the western Pacific Ocean. To be even more specific, the waters within a triangle formed by the Philippines, Malay Peninsula and New Guinea are considered the richest in terms of tropical marine species diversity, both for fish and invertebrates (especially the latter). Over millions of years this large region has been subject to a stable, warm, tropical climate, exactly the kind of physical factors that encourage the diversification of species. Here competition for food, space and mates has been the primary force driving the natural selection engine.

The marine life radiating from this region is known as the Indo-West Pacific shore fauna, and it is the largest marine-faunal area in the world. In this case, a faunal area is defined by the kinds of similar organisms you find in a particular geographic location. The Indo-West Pacific faunal region extends halfway around the globe, from the Red Sea and the eastern coast of Africa eastward across the entire tropical Indian Ocean to the western Pacific Ocean. It does not extend across the east Pacific to North America. In the northeastern Pacific, it extends no further

Male and female goldfish look significantly different . This male displays a characteristically high first dorsal fin ray, bright colors and larger size compared to the yellow-orange females surrounding him.

FAGATELE BAY

Habitat
Tropical coral reef

Key Species
Crown-of-thorns starfish
Blacktip reef shark
Surgeon fish
Hawksbill turtle
Parrotfish
Giant clam

Description
Fagatele Bay comprises a fringing coral-reef ecosystem nestled within an eroded volcanic crater on the island of Tutuila. Nearly 200 species of coral are recovering from a devastating crown-of-thorns starfish attack in the late 1970s, which destroyed over 90 percent of the corals. Since then, new growth has been compromised by two hurricanes, several tropical storms and coral bleaching. This cycle of growth and destruction is typical of tropical marine ecosystems.

Cultural Resources
3,000+-year-old thriving Polynesian culture originated in Samoa

Location
The southwest shore of Tutuila Island, American Samoa, 14 degrees south of the equator

Protected Area
.25 square miles

Like a sand bulldozer, this foot-long nudibranch is cruising along at night in a sand channel on Maui, searching for small snails, worms and small clams.

east than Hawaii. In the southeastern Pacific, it reaches all the way to the Tuamoto archipelago having its easternmost boundary at Hawaii. What this really means is that the center of genetic material for the great majority of marine life within this region originated from the triangle of water described above. Even the marine life of the Tuamotos has its beginnings in the shrimp, worms, clams, fishes and other marine life whose larvae somehow crossed thousands of miles of water to escape the intense competition of the Indo-West Pacific heartland. To put it in perspective, today there are approximately 2,000 species of reef and shorefishes in the Philippines, about 1,000 species in the Marshall Islands and about 450 species in Hawaii. As the distance from the center of diversification increases, the number of species that can successfully invade and establish new populations in other similar habitats decreases.

In a sense, Hawaii is at the end of the line of this long migration eastward.

The great dispersal of marine life across the tropical Pacific could not have taken place had it not been for two things. First, many, if not most, marine invertebrates and reef fishes produce young in huge numbers that greet the world as small, planktonic larvae. Second, these young are easily dispersed in water which both keeps them afloat and transports them—often great distances—in the ocean currents. How long they are able to survive floating around in the sea depends on the length of their larval stages (which vary greatly from species to species), their chances of being eaten (generally speaking, the longer they are floating around, the better their chances of becoming someone's meal) and on how long it takes to find a suitable environment to colonize. Clearly, these last two factors are at odds with each other. On the one hand, the longer you are traveling the better the chances of finding (perhaps even selecting) a good place to live. However, the chances of being devoured also improve with extended periods as a plankter. How far and in what direction a larva floats is determined by the strength and direction of the current carrying it. Clearly, it is

This egg mass of the Spanish dancer nudibranch, or sea slug, contains thousands of tiny eggs.

advantageous to have as many young as you can to improve the chances of a few making a trip to another location where competition with each other and other species may be less. Through a process of elimination, the species originating in the Indo-West Pacific region that were best adapted for long periods of "hang time" (travel time as a planktonic larva) and could survive in the subtropical waters of Hawaii are also the species that successfully invaded and established themselves in remote places like Hawaii.

Top: At the sea's edge, ghost crabs dig extensive burrows from which they emerge at night on feeding forays. This one was caught out just before sunrise on a beach on Maui.

Below: Swimming crabs use their large, flattened hind legs for sculling through the water sideways. This one is taking a late night swim through the waters of Maui.

Taking a rest from feeding on plankton, this clownfish is "communicating" with its host anemone by rubbing against it. The two animals recognize each other from the unique chemical make-up of their skins.

Island Hopping

While some species of invertebrates and fish have larvae that remain in a planktonic stage for weeks, even these could not make the trip from the Philippines all the way to Hawaii—over 3,000 miles. Like the Polynesians, it would appear that tropical fishes and invertebrates having long-lived planktonic larvae adopted the same island-hopping strategy, working their way east one island group at a time. Where islands or even submerged reefs are grouped in clusters, the distances between them are not so great as to preclude colonization by even those species having medium-length planktonic lives. Local water patterns, currents and eddies move these young animals and plants from coral head to coral head, reef to reef and so on. In this way colonization takes place in a step-wise fashion. As each new island group is reached, a healthy population is established which reaches a level that "broadcasts" enough larvae so that statistically a few will make it to the next "landfall." Keep in mind, these small organisms are not planning their travels. Success or failure in finding a new settling site is often the result of simply being in the right place at the right time. However, those

Scuttling along a shallow reef, this anemone crab gingerly carries two companions in each claw. The anemones help defend the crab, and in return they are carried to more food sources than stationary anemones.

Goldfish (not the freshwater, pet store variety) are common schooling fish of Indo-Pacific reefs. They hover in groups over coral heads, darting at small plankton passing in the current.

species that have extended planktonic stages obviously have an adaptive advantage in staking out new territory and escaping the competitive environment from which they were conceived.

While islands and shallow reefs serve as way stations for the distribution of marine life in the tropical Pacific, conversely, large expanses of open ocean act as barriers. Migration and colonization of the eastern portions of Micronesia seem to have been relatively successful for a number of species originating farther west. This may be due in large part because of the absence of any great open ocean areas such as the 1,000+ miles separating Hawaii from the Marshall Islands, the nearest island chain to the west. South of the Equator there is a nearly continuous band of islands extending from Indonesia to Tahiti. While Tahiti is approximately the same distance from the Equator as Hawaii, it has many more species with clear Indo-West Pacific affinities. It appears that

Peering sheepishly out of a hole in the reef, this small blenny is typical of the thousands of small animals that live deep in the reefs.

There is no safer place for a juvenile clownfish than tucked among the protective stinging tentacles of a carpet anemone.

Johnston Island, 450 miles south of Hawaii, has played a significant role in the colonization of Hawaii in that its marine life shows strong faunal relationships with Hawaii. In fact, it may be that some species that successfully populated Hawaii from Johnston Island left Hawaii and repopulated Johnston, creating future generations of species whose genetic codes have made a round-trip between the two island groups.

But the greatest open ocean barrier still seems to be the stretch between Hawaii and the North American continent. Very few species have made it all the way across the Pacific. In all, about twenty to thirty Indo-West Pacific fishes have been found along the shorelines of Mexico, Central and South America, but almost none have made it across the Pacific from east to west.

These large, expansive plate corals provide shelter for a variety of small, shallow-reef organisms from crabs to small fish.

Animals Found Nowhere Else in the World

One of the most characteristic patterns to develop in isolated environments is a high degree of endemism, and this is certainly the case in Hawaii. For example, approximately 30 percent of the shallow-water sea stars and brittlestars, 45 percent of the shrimp family *Crangonidae*, and 20 percent of Hawaiian molluscs are found nowhere else in the world. The inshore fishes show a particularly high degree of endemism: 34 percent. Some of the most familiar and common Hawaiian reef fishes are endemics. The ubiquitous saddleback wrasse and pedestrian manini (a common surgeonfish) are both endemic species. In some instances, both the "parent" form of the fish and the endemic form that evolved from it are currently found

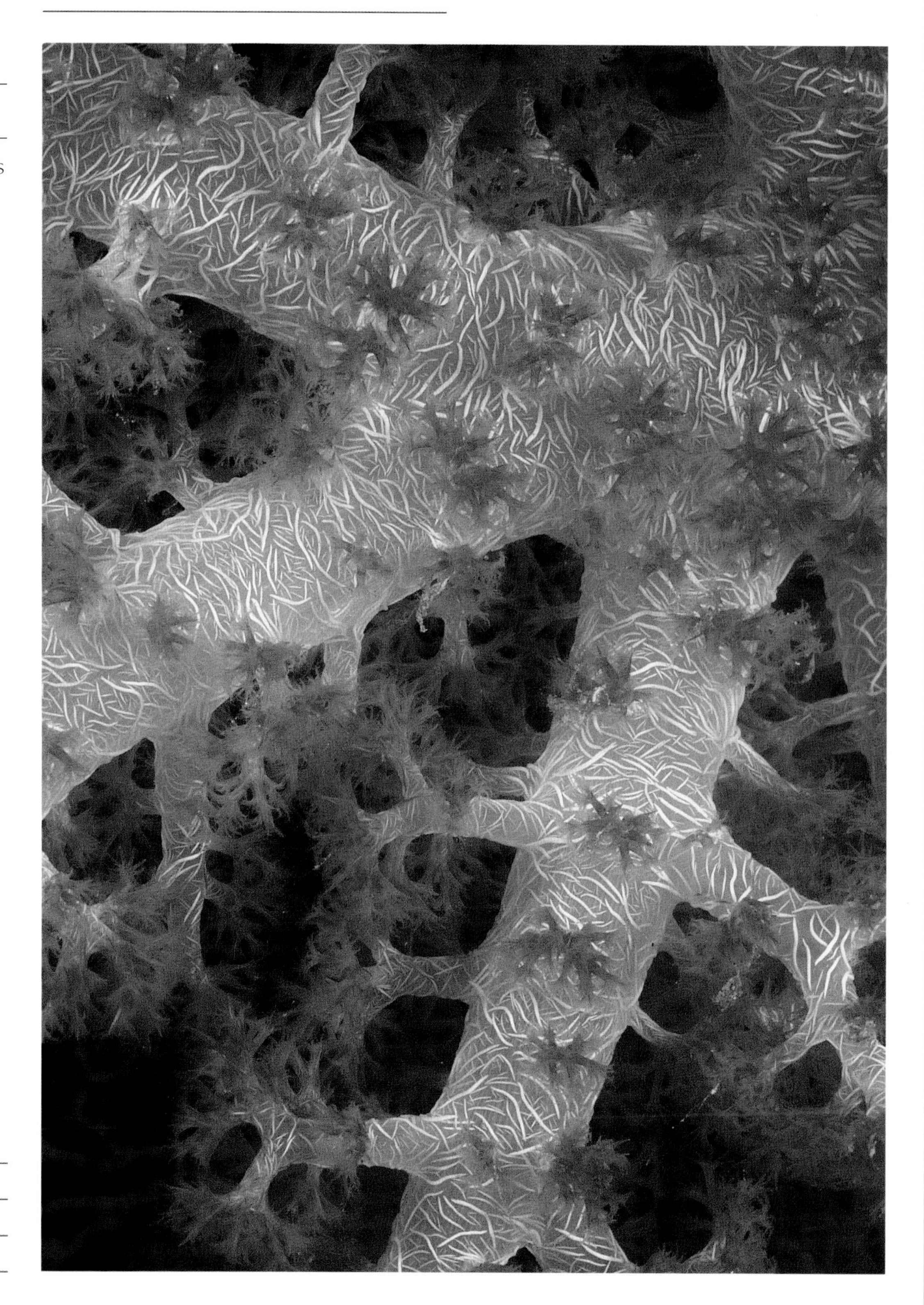

Soft corals are members of the octocoral group, each polyp having eight distinct tentacles. They are common members of Indo-Pacific reefs.

in Hawaiian waters. This seems to be the case for the lizardfishes and the cardinalfishes. The theory is that the parent form re-entered the Hawaiian region (possibly from nearby Johnston Island) after the original parent had disappeared by evolving into the Hawaiian endemic version.

Two aspects of Hawaii's geographical location seem to be primarily responsible for these large percentages of endemic species. First, because of its isolation the organisms that have migrated to Hawaii can only interbreed within the few individuals that successfully made the journey. A small gene pool greatly improves the odds of new species being formed. A morphological or behavioral adaptation that gives an individual an advantage can quickly spread through a small population because there is a limited number of like individuals with which to breed. As in other isolated environments, Hawaiian fishes and invertebrates have had an unusual opportunity to evolve into new

Unlike its cousin the porcupinefish, the black-spotted puffer does not erect sharp spines when it inflates itself in defense. It depends solely on becoming a difficult mouthful.

One of the world1s largest pufferfish, this species may grow to nearly three feet in length.

Tucked back in a hole by day, this moray eel awaits nightfall, when it will emerge from this Hawaiian reef in search of unsuspecting nocturnal prey.

and unique species as a result of this isolation. Additionally, because there is (and has been) little pressure from outside forces (such as other competing species) those species that develop particularly useful adaptations can become very successful, in terms of population, in a short period of time.

The second contributing factor to high endemism related to geography is Hawaii's subtropical climate. Hawaii has always enjoyed a mild and stable climate, an environment that is also conducive to species diversification in general. This is the same theory that has been applied to the general evolution of marine life for the entire Indo-West Pacific region. Namely, that the combination of a stable climate over an enormous area should encourage new species. Without a rigorous physical environment to deal with, the environment does not play as great a role in shaping species diversity. Thus, endemic species have evolved from parent forms in order to reduce competition for resources and to take advantage of other ecological niches. This might mean feeding on something that no other related species eat, or taking advantage of microhabitats that their brethren do not inhabit. It may also partially explain how certain species can disperse their progeny over large geographic areas having warm, stable climates.

But as mentioned earlier, Hawaii is really at the northernmost latitude for coral growth and is considered more subtropical than tropical. Cooler waters have in part helped create a marine community that is truly unique to Hawaii.

Displaying its venomous dorsal spines, this lionfish patrols a shallow reef near Fiji. The spines are used for defensive purposes only.

In addition to great endemism, Hawaiian marine life is subject to a certain degree of what biogeographers call disharmony. Some species that are very common in Hawaii are not so common in other parts of the Indo-Pacific range. A good example here is the slate pencil sea urchin, which is extremely abundant in Hawaii but not in other tropical reef areas. The saddleback wrasse has already been mentioned as an endemic, and is also present in Hawaii in great numbers. Species that are common in most other parts of the Indo-West Pacific region but absent, or nearly so, in Hawaii include corals in the family *Acroporidae*, and fishes in the snapper and grouper families. Is the lack of distribution of these animals in Hawaii simply luck of the draw in terms of riding the right current, or is it because they don't have the "right stuff" genetically to make it in Hawaii?

On a shallow Hawaiian reef called Ulua, this juvenile anglerfish rests peacefully in a small coral head.

Shy and retiring, falco hawkfish dart in and out of the reef framework. The small tufts of bristles on the tips of the dorsal spines are characteristic of this species.

Subtidal Habitats

Exploring Hawaii's underwater wilderness begins at the shore, where anyone with mask, fins and snorkel can comfortably wade into the surf and enter a world filled with colorful tropical fish, corals and sunlight. Because of its warm subtropical climate Hawaii's marine life is accessible to nearly everyone.

As in the Florida Keys, Hawaiian corals grow best in quiet, well-lit, relatively warm water. In very protected areas, such as Manele-Hulopoe on Lanai, coral develops profusely only 10 or 15 feet below the surface. However, in most parts of Hawaii, wave action plays a critical role in the distribution of corals, with most living below 25 to 35 feet to escape the surge of powerful waves rolling in overhead. These waves and swells build as winds blow over the great expanse of the Pacific Ocean. These ocean swells hit Hawaii from many directions, but there is a seasonal correlation, with the trades creating the windward and leeward sides of the islands described above. At any given location one type of water motion usually predominates. In protected environments such as bays and lagoons, wave surge is slight whereas along the open coast wave surge predominates over weaker wind-driven and tidal currents. In shallow water the surge and turbulence associated with large waves can damage corals and the entire community of other organisms living on and in it. The most violent waves along Hawaiian coasts are those originating from winter storms coming out of the North Pacific. Generally speaking, the

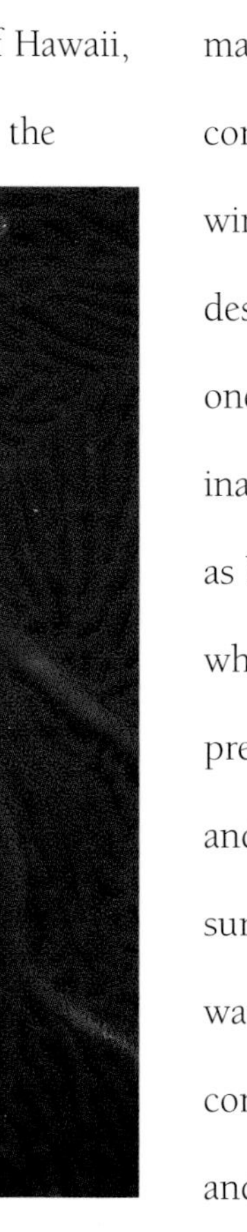

A small goby rests on the primary "branch" of a sea fan in the Indo-Pacific. The goby's coloration matches the background of the sea fan, which probably accounts for its entire world view.

most diverse coral growth is found in water shallower than 30 to 40 feet on the leeward side of islands, such as on the Kona Coast of the Big Island. Wave action is perhaps the primary physical factor determining the distribution of corals and their associated marine life throughout the islands.

Hawaiian Coral Communities

Being located at the northern edge of the coral reef zone, Hawaiian corals live within the lower end of the temperature range necessary for reef-building corals to grow. While this means that Hawaiian corals are not as diverse as the reefs of other Pacific islands closer to the Equator and Indonesia, they are still present in numbers that allow for the creation of legitimate coral reef structures. Forty different kinds of reef-building corals are found in Hawaii, while a few of the more spectacular and conspicuous octocorals are practically absent altogether. This is perhaps one of the most outstanding

These Favia coral polyps appear soft and mushed together, but, in reality, they are made from hard calcium carbonate like all other reef-building corals.

Like two peas in a pod, this pair of lizardfish seem perfectly happy to snuggle up to each other in a sand channel off Maui.

differences between the reefs in the heart of the Indo-West Pacific region and Hawaii. There are no colorful, branching soft corals, no enormous sea fans, nor are there any candelabra-shaped sea whips. Rather, stony corals dominate the reef landscape, which already appears hardened by the exposed outcrops of dark basaltic lava.

Corals play a particularly crucial role in reef building because they provide a large part of the actual physical structure of the reef, modifying the surrounding seascape with the addition of their large, multi-faceted skeletons. As such, corals determine to a large degree the community structure of a reef, including the biological assemblage of other organisms—both plant and animal—in terms of species diversity and abundance. Corals create an irregular, three-dimensional structure creating shade, protection from surge, accumulations of sediments and numerous "microhabitats" such as cracks, crevices, nooks and crannies within which a diversity of organisms may live and flourish.

Illuminated with late-day sun, this jelly was photographed in a shallow lagoon near Fiji.

As with Caribbean reef-building corals, the species in Hawaii responsible for reef growth rely on symbiotic zooxanthellae algae within their tissues. In many of Hawaii's shallow-water environments that are particularly exposed to wave surge, the stirring of sediments may create such turbid water as to critically reduce the ability of zooxanthellae to photosynthesize, thus slowing the coral's ability to grow. The relatively quiet water below the surge zone allows adequate light penetration for these corals and their symbiotic

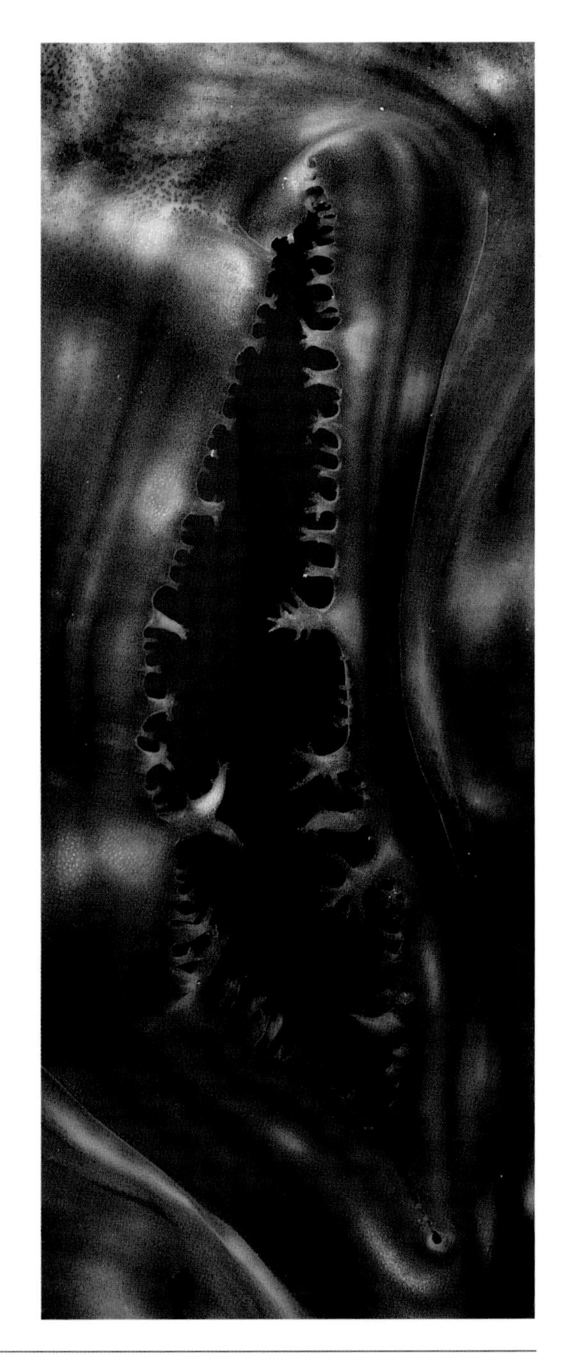

This extreme close-up of the intake valve of a giant clam shows the mottled colors created by the uneven distribution of symbiotic algae living in the tissues of the clam's mantle.

algae to vigorously grow. At this depth they are also protected from the constant abrasion caused by the movement of sand particles brushing against live corals. The Kona coast has perhaps some of the most extensive and healthy stands of corals anywhere in the islands, thanks to its relatively protected location from the trade winds and winter storm swells. Coral can cover as much as 20 percent of the bottom even in the shallows along the Kona coast, while in waters 30 to 70 feet, coral cover can be 70 percent.

Although corals are major contributors to Hawaiian reefs, many other encrusting organisms also play important roles in creating the reef's physical structure. Coralline algae are the main structural builders of reefs, especially in surge-swept areas where turbidity may be too high to allow for significant coral growth. This is also the case on the outer-most wave-swept portions of a reef. This area of encrusting algae is called the algal ridge and it is located at the front of the reef that is exposed to the greatest wave action. Encrusting algae not only provides a solid substrate, it also acts as the mortar of the reef, binding together the solid and loose fragments of limestone sediments swirling around coral heads.

Hawaiian reefs do not normally approach or grow above low tide where they would be exposed to the island's relatively extreme subtropical climate. This is also partly due to the absence of Acroporid corals, a family that is particularly well-adapted to withstanding periods of exposure to the air. In Hawaii, reef growth in any particular direction depends on physical factors such as depth, turbulence, clarity, sediment movement, temperature and salinity. Areas scoured by sand, buried under mud or diluted by freshwater runoff are generally not conducive to reef building.

Even where conditions for growth are

Who would have thought that the armpit of a lionfish would reveal such visual beauty and detail?

most favorable, such as on the Kona coast, physical and biological erosion breaks down the reef. Physical erosion comes mainly in the form of wave action. Biological erosion is due to boring organisms such as worms, sponges, clams, crabs, sea urchins and others in their attempt to make a comfortable home for themselves in the protection of the reef. Coral grazing fishes, such as parrotfish, also contribute significantly to reef erosion. At first, any eroded material will fall to the base of the reef. From here it will eventually be carried into deeper water, or during periods of intense wave or surge action it may be transported over or through the reef, contributing to a sandy beach behind it.

As the physical material of the reef is being eroded, it is replaced by other reef-building organisms. In a balanced system, the net result should be one of equilibrium over the long-term. Of course, each situation and reef location in Hawaii is differ-

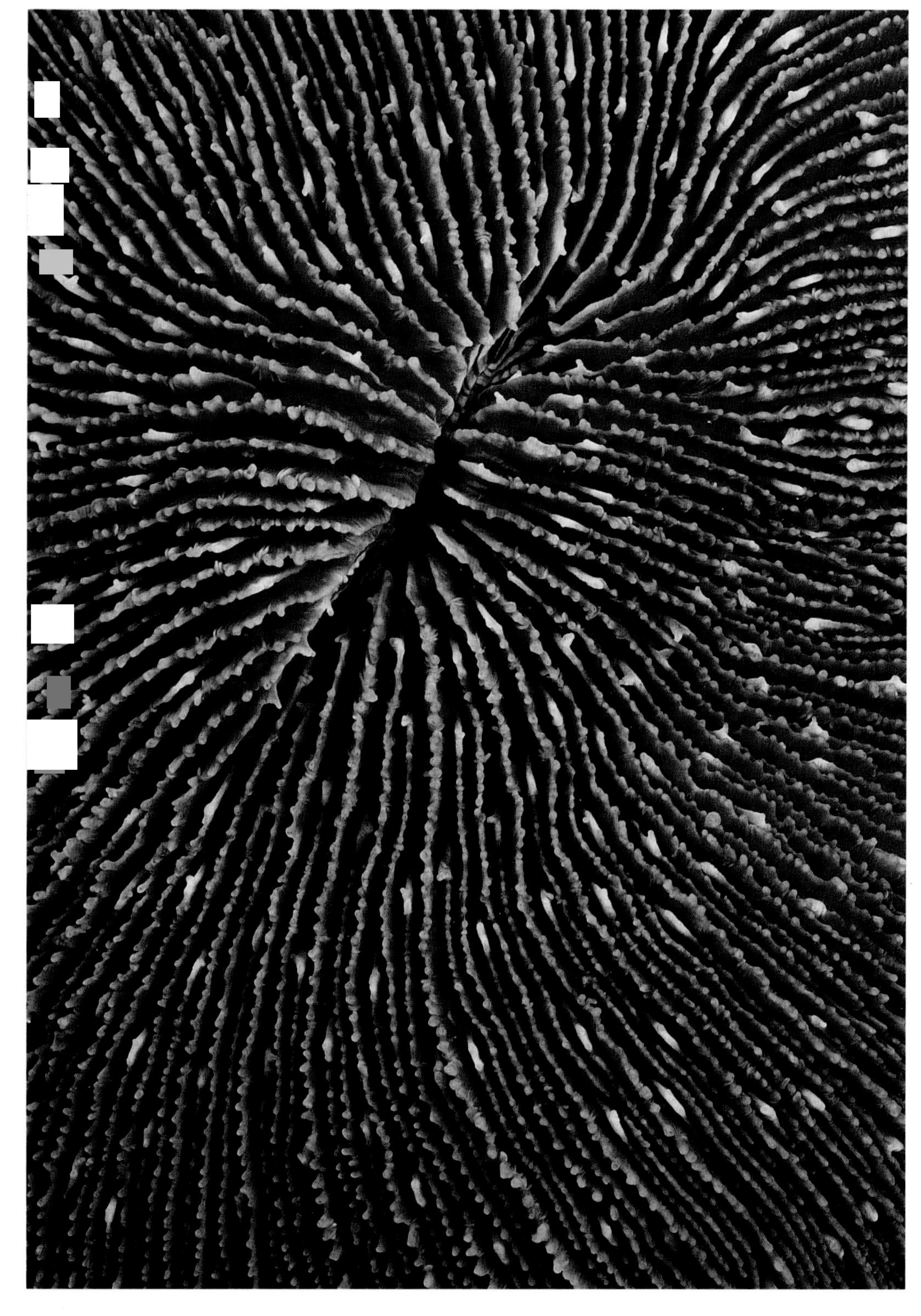

ent, but in general the windward reefs of many of the islands appear to be eroding as fast or faster than they are growing. The opposite is true for many reefs located on leeward shores. Where growth exceeds erosion, a reef will gradually extend upwards and outwards. This process is an agonizingly slow one given the growth rates of certain corals—about half an inch per year under ideal conditions. Contrast this to the relatively fast growth rates seen in the Caribbean and places in the tropical

"All grooves lead to the mouth" in this extreme close-up of a mushroom coral.

Pacific where Acroporid corals are present and abundant. These species may grow several inches a year under ideal conditions. Given all this, the fragile nature of coral reefs becomes apparent, knowing that their growth depends directly on the health of the corals, coralline algae and other encrusting organisms that make up this relatively thin layer of living tissue on the outer reef flat and reef front.

Detail of reef anemone, Indo-Pacific.

Appearing to be floating in air, this squid is hovering just below the water's surface over an Indo-Pacific reef.

Glossary

Bathymetry. The study and mapping of sea floor topography.

Benthic. Organisms that live permanently on or in the bottom of the ocean.

Conduction. The transfer of energy through matter by internal particle or molecular motion without any external motion.

Continental margin. The edge of a continent.

Convection. Vertical circulation caused by density differences within a fluid, resulting in a transport and mixing of the properties of that fluid.

Convection cells. Convective fluid movement in a mass, with a central portion moving upward and outer regions moving downward. Convection cells in the asthenosphere move lithospheric plates horizontally across the surface of the earth.

Errant. Moving freely about.

Eurythermic. An organism capable of withstanding a wide range of salinities.

Eustatic. Pertaining to worldwide and simultaneous change in sea level, such as that caused by melting of glaciers.

Foraminifera. An order of protozoans, mostly benthic or living attached to plants and animals, having internal chambered shells usually composed of calcite or cemented sand grains.

Island arc. A linear or arcuate chain of volcanic islands formed at a convergent plate boundary.

Longshore current. A current in the surf zone, moving parallel to the shore and generated by waves breaking at an angle to the beach.

Magma. Molten rock in the earth's crust or mantle that crystallizes to form an igneous rock.

Neap tide. The lowest tidal range during the lunar cycle, when the sun and moon are at right angles to the earth. It occurs twice each month, at the first and third quarters of the moon.

Plate tectonics. The theory and study of plate formation, movement, interactions and destruction; the attempt to explain seismicity, volcanism, mountain-building and paleomagnetic evidence in terms of plate motions.

Pleopods. Abdominal-paired appendages in crustaceans.

Pressure cell. A portion of the atmosphere that behaves as a unit.

Relic. In geology, a rock formation left behind after the decay, disintegration or disappearance of surrounding geologic features.

Sessile. Permanently attached to a substrate such as the sea floor.

Shield volcano. A large, broad volcanic cone with very gentle slopes built up by nonviscous basalt lavas.

Shoal. A shallow sandbank or sandbar.

Soundings. A measurement to ascertain the depth of water.

Swimmeret. One of a series of small, unspecialized appendages under the abdomen of many crustaceans, often used for swimming and carrying eggs.

Telson. The terminal segment of the body of an arthropod or segmented worm.

Terrigenous. Being or relating to oceanic sediment derived directly from the destruction of rocks on the earth's surface.

Zoogeography. The study of the geographical distribution of animals.

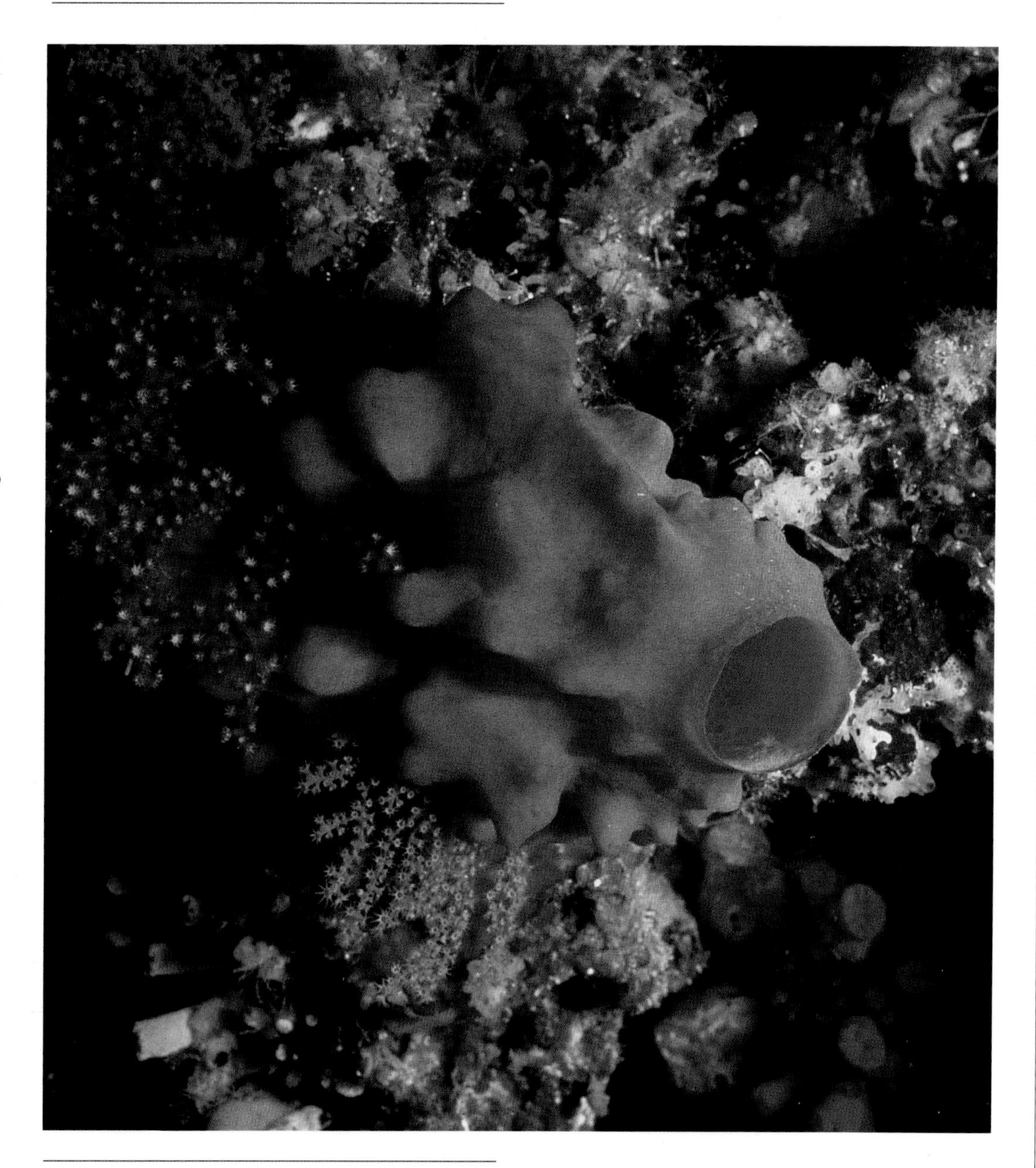

Like an underwater still life, this collection of yellow sponges, sea fans, corals and sea squirts is a typical section of many reefs throughout the tropical Pacific.

National Marine Sanctuaries & Reserves

NATIONAL MARINE SANCTUARIES

Channel Islands National Marine Sanctuary, California
Tel: (805) 966-7107

Cordell Bank National Marine Sanctuary, California
Tel: (415) 561-6622

Fagatele Bay National Marine Sanctuary, American Samoa
Tel: (684) 633-7354

Florida Keys National Marine Sanctuary, Florida
Tel: (305) 743-2437

Flower Garden Banks National Marine Sanctuary, Texas
Tel: (409) 847-9296

Gray's Reef National Marine Sanctuary, Georgia
Tel: (912) 598-2345

Gulf of the Farallons National Marine Sanctuary, California
Tel: (415) 561-6622

Hawaiian Islands Humpback Whale National Marine Sanctuary, Hawaii
Tel: (808) 879-2818

***Monitor* National Marine Sanctuary, Virginia**
Tel: (804) 599-3122

Monterey Bay National Marine Sanctuary, California
Tel: (408) 647-4201

Olympic Coast National Marine Sanctuary, Washington
Tel: (360) 457-6622

Stellwagen Bank National Marine Sanctuary, Massachusetts
Tel: (508) 747-1691

NATIONAL ESTUARINE RESEARCH RESERVES

ACE Basin National Estuarine Research Reserve, South Carolina
Tel: (803) 762-5062

Apalachicola Bay National Estuarine Research Reserve, Florida
Tel: (904) 653-8063

Chesapeake Bay National Estuarine Research Reserve in Maryland
Tel: (410) 974-3382

Chesapeake Bay National Estuarine Research Reserve in Virginia
Tel: (804) 642-7135

Delaware National Estuarine Research Reserve, Delaware
Tel: (302) 739-3451

Elkhorn Slough National Estuarine Research Reserve, California
Tel: (408) 728-2822

Great Bay National Estuarine Research Reserve, New Hampshire
Tel: (603) 868-1095

Hudson River National Estuarine Research Reserve, New York
Tel: (914) 758-5193

Jobos Bay National Estuarine Research Reserve, Puerto Rico
Tel: (787) 853-4617

Narragansett Bay National Estuarine Research Reserve, Rhode Island
Tel: (401) 683-5061

North Carolina National Estuarine Research Reserve, North Carolina
Tel: (910) 256-3721

North Inlet/Winyah Bay National Estuarine Research Reserve, South Carolina
Tel: (803) 546-3623

Old Woman Creek National Estuarine Research Reserve, Ohio
Tel: (419) 433-4601

Padilla Bay National Estuarine Research Reserve, Washington
Tel: (360) 428-1558

Rookery Bay National Estuarine Research Reserve, Florida
Tel: (941) 417-6310

Sapelo Island National Estuarine Research Reserve, Georgia
Tel: (912) 485-2251

South Slough National Estuarine Research Reserve, Oregon
Tel: (541) 888-5558

Tijuana River National Estuarine Research Reserve, California
Tel: (619) 575-3613

Waquoit Bay National Estuarine Research Reserve, Massachusetts
Tel: (508) 457-0495

Weeks Bay National Estuarine Research Reserve, Alabama
Tel: (334) 928-9792

Wells National Estuarine Research Reserve, Maine
Tel: (207) 646-1555

Cup corals are night-active predators of small planktonic organisms. By day their large, brilliant orange-red tentacles are withdrawn, emerging at dusk to begin feeding through the night.